Current Sensing Techniques and Biasing Methods for Smart Power Drivers

Sri Navaneethakrishnan Easwaran

Current Sensing Techniques and Biasing Methods for Smart Power Drivers

 Springer

Sri Navaneethakrishnan Easwaran
Texas Instruments (United States)
Dallas, TX, USA

ISBN 978-3-319-89127-9   ISBN 978-3-319-71982-5   (eBook)
https://doi.org/10.1007/978-3-319-71982-5

Printed on acid-free paper

This Springer imprint is published by Springer Nature
The registered company is Springer International Publishing AG
The registered company address is: Gewerbestrasse 11, 6330 Cham, Switzerland

With tons of thanks to my supervisor
Prof. Dr. R. Weigel, Friedrich-Alexander-
Universität Erlangen-Nürnberg.

Abstract

The demand for electronic control is increasing in the automotive industry and is quickly replacing hydraulics. Automobiles produced today have a longer life expectancy, and one of the strongest contributors to this is how electronic equipment and systems have replaced mechanical devices. Without proper electrical protection, however, these electronic systems in automobiles can fail without warning. This creates a risk for both the driver and passengers, including injury and death. Power semiconductor devices that are used in these electronic systems handle high currents in the order of 2 A at 40 V. Due to this high voltage and high current, several design measures have to be considered when designing how to make these circuits fault tolerant. The unintentional or inadvertent activation of the powerFETs should be avoided to prevent very high currents. This book contributes to the design of smart power drivers for airbag squib driver applications. It focuses on the current regulation circuits with fault-tolerant topologies and inadvertent current control. Inadvertent current flows can damage safety ASICs or even alter safety functionality. The designs have to be developed to withstand faults like missing power supply in a Multiple Supply Domain ASIC that can lead to high currents due to floating nodes. The smart power drivers should be guaranteed to not misfire the airbags when some supplies are missing. Other faults that can lead to airbag misfire are short-to-battery and short-to-ground conditions, which happen during either the powered or unpowered state of the ASIC.

This book extends the technical state of the art and improves on it by the following principle results:

1. The current sensing scheme is designed to achieve the desired current limitation in the high-side configuration with a metal sense resistor. In addition to the current regulation, the novel surge current control circuit to limit the current through the powerFET is needed to limit the duration of the peak current. The importance of this is explained and verified. The duration is 5 A for 4 μs.
2. A current sensing scheme meant to achieve the desired current limitation in the low-side configuration with a senseFET-based topology is presented. The novel

surge current control circuit is meant to ensure that the circuit is fault tolerant when a short-circuit-to battery (SCB) condition occurs. The duration is 5 A for 4 μs.
3. The exploitation of current selector circuits to ensure well-defined bias conditions for the powerFETs and circuits operating with multiple supply voltages (MSVs) is discussed.
4. A diagnostic section is discussed with two specific targets. Diagnostic circuits consist of a voltage reference circuit and an output buffer. The design of a precise current limited voltage source (CLVS) is discussed. The common mode input of the output buffer can range from 0 to 40 V, but the output level of this buffer is limited to 5.5 V by an improved clamping technique.

A four-channel squib driver prototype has been manufactured on a 350 nm, 40 V BiCMOS process. Extensive measurements have been performed for the current sensing circuits on both high-side and low-side configurations. The high-side current sensing scheme regulates to 2 A with ±3% accuracy. The low-side configuration regulates to 3 A with ±10% accuracy. The peak current limitation in the unpowered state is 1 A for 400 ns and 3 A for 1.5 μs for the high-side and low-side configurations, respectively. The high-side and low-side drivers have stable current regulation for inductive loads up to 3 mH. High-side drivers provide an internal freewheeling path during turn off conditions. The biasing scheme introduced with voltage and current selectors did not show any unintended activation or deactivation of the circuits under any power-up condition. Current limited voltage source for diagnostics implemented in both high-side and low-side configurations regulate to 8.4 V with ±0.5 V accuracy. Output buffers regulate to 5.4 V with ±200 mV accuracy under any power-up phase of operation.

Acknowledgements

This book is the result of my Doctor of Philosophy assignment at Texas Instruments Inc., Dallas, Texas, USA. The project was carried out from January 2010 to August 2016 at the Mixed Signal Automotive (MSA) group at Texas Instruments Inc. It was used to complete my Doctor of Philosophy in Electrical Engineering in the Department Elektrotechnik at the Universität Erlangen-Nürnberg.

I would like to thank Mr. Joseph A. Devore, product line manager, Integrated Automotive Products (IAP), for allowing me to lead the airbag squib driver development project, which has been instrumental for this dissertation. He helped me in my career at Texas Instruments Inc., and supported me while I wrote my thesis. I would like to thank Mr. Sigfredo Gonzalez Diaz, engineering manager, IAP, and my supervisor Mr. Benjamin Amey for their encouragement and support on this project.

I would like to convey my deepest gratitude to Prof. Dr.-Ing. habil. Robert Weigel for taking me on as his external PhD candidate. He guided me towards the completion of my thesis. His support was exceptional, and I appreciate the individualized attention he gave to me, taking the time to clarify my queries and correct my mistakes even though we were located on two different continents. As my supervisor, he constantly pushed me to achieve my goals. His observations and comments helped me to establish the overall direction of the research and move forward with the investigation. I would like to thank Prof. Dr.-Ing. Martin März for all his support as well. He gladly accepted to be one of my reviewers and provided additional, helpful feedback to improve my thesis.

I greatly appreciate Dr. Shanmuganand Chellamuthu at Texas Instruments Inc., Dallas, Texas, USA, for all his guidance, too. His support was exceptional, especially in how he helped me systematically organize my thesis. I would like to express my gratitude to Dr. Samir Camdzic and Dr. Ralf Brederlow, Texas Instruments GmbH, Freising, Germany, as well for their constructive feedback. I am also grateful and thankful to Dr. Krishnaswamy Nagaraj (TI Fellow) and Dr. Jingwei Xu (TI Fellow) for their reviews and valuable feedback. I have no words to express my gratitude for the support from Mr. Larry Bassuk and Mr. Bill Kempler, Legal

Department, Texas Instruments Inc., Dallas, Texas, USA, for helping me publish my thesis.

I thank Prof. Dr. Chandra Sekhar, CEERI, Pilani, India, Prof. Dr. ir. M.J.M. Pelgrom and Prof. Dr. ir. B. Nauta and his staff, ICD, University of Twente, Enschede, The Netherlands, for being a source of inspiration. I would like to thank several of my professors and airbag control unit experts who have been a significant source of knowledge and inspiration for my research.

I would like to thank my colleague Mr. Sunil Kashyap Venugopal, Texas Instruments Inc., Dallas, Texas, USA, for providing me the digital design support. I thank my parents for their support and blessings, my wife Hemalatha and daughter Vaisshnavi for their unconditional support and encouragement, and my brother Sai for his support.

Contents

Contents   xiii

Table of Symbols

g_m	Transconductance in siemens (S)
g_d	Output conductance in siemens (S)
V_{BAT}	Battery voltage in volt (V)
R	Output impedance in ohm (Ω)
$A(s)$	Open-loop gain in decibel (dB)
β	Feedback factor
ϕ_m	Phase margin in degrees ($^\circ$)
C_L	Load capacitor in farad (F)
f_p	Frequency of the pole in hertz (Hz)
f_z	Frequency of the zero in hertz (Hz)
f_τ	Unity gain bandwidth in hertz (Hz)
μ_n	Mobility of the electron in m^2/Vsec
C_{ox}	Oxide capacitance of the MOS transistor in farad (F)
C_{gs}	Gate to source capacitance of the MOS transistor in farad (F)
C_{gd}	Gate to Drain capacitance of the MOS transistor in farad (F)
T_j	Junction temperature in $^\circ$C
T_{crit}	Critical temperature of powerFETs in $^\circ$C
θ_{JA}	Thermal resistance from junction to ambient C/W
σ_{vt}	Standard deviation of the threshold voltage in mV
R_{ds_on}	Channel *on* resistance of the FET in ohms
V_{th}	Threshold voltage in volt (V)
μC	Microcontroller

Abbreviations

ACU	Airbag control unit
ADC	Analog to digital converter
CLVS	Current limited voltage source
HS_FET	High-side field effect transistor
HV	High voltage
LDMOS	Laterally diffused metal oxide semiconductor
LDO	Low dropout regulator
LHP	Left half-plane
LS_FET	Low-side field effect transistor
LV	Low voltage
MCS	Maximum or minimum current selector
MCU	Microcontroller
MSV	Multiple supply voltage
MV	Medium voltage
OTA	Operational transconductance amplifier
PMU	Power Management Unit
RHP	Right half-plane
SBC	System Basis Chip
SCB	Short circuit to battery
SCG	Short circuit to ground
SDU	Squib Driver Unit
SOA	Safe operating area
UGB	Unity gain bandwidth

About the Author

Sri Navaneethakrishnan Easwaran was born in Erode, India, on October 19, 1977. He received his Bachelor of Engineering, B.E., degree (cum laude) in Electronics and Communication Engineering from Shanmugha College of Engineering (affiliated to Bharathidasan University, Tiruchirapalli, India), Thanjavur, India, in 1998. He worked at SPIC Electronics, Chennai, and STMicroelectronics, Noida, India, between 1998 and 2000. From 2000 to 2006, he worked for Philips Semiconductors at Bengaluru, India; Zurich, Switzerland; Nijmegen, The Netherlands where he designed analog circuits for the Mobile Baseband and Power Management Units. While working at Philips Semiconductors, he received an international M.Sc. degree in Electrical Engineering from the University of Twente, Enschede, The Netherlands on the design of NMOS LDOs. Beginning in 2006, he started his work at Texas Instruments GmbH, Freising, Germany. He joined the Technische Elektronik group at Friedrich-Alexander-Universität Erlangen-Nürnberg in January 2010 as an external PhD student. His research focused on the fault-tolerant design of smart power drivers and diagnostic circuits. He received his Dr.-Ing degree from Friedrich-Alexander-Universität Erlangen-Nürnberg in May 2017. Since September 2010, he has worked with Texas Instruments Inc., Dallas, Texas, USA. He was elected as a Senior Member of IEEE in 2011 and part of the Member Group Technical Staff at Texas Instruments in 2014. He has more than 15 patents (USA and German) in the field of Analog Mixed Signal IC Design and has 6 IEEE and conference publications.

Chapter 1
Introduction

Ever since the microprocessors got introduced in electronics, the electronic controls have gained immense use in household appliances to large-scale complex industrial controls. In the last two decades, the need for electronic controls along with the development of power semiconductor devices like LDMOS (laterally diffused metal oxide semiconductor) has shown aggressive growth in the integrated power devices along with their driver circuitry. For example, integrating 40 V rated, low resistance transistors (300 mΩ) for switches, regulators, current sensing, etc. along with their control/driver circuitry is a state of art. This led to the miniaturization of the PCB (printed circuit board) space due to less external components paving its way into maximizing the electronic content in automotive. Today, electronics represent around 25% to 30% of a modern automotive. The demand for electronic control is increasing and is replacing most of the hydraulics in the automotive. As shown in Fig. 1.1, between 2012 and 2017, the automotive electronic production has grown and is expected to grow at an average annual rate of 7.0% worldwide, to reach 179 billion euros by the end of the period [1]. The whole of the automotive industry rely on electronic systems and technologies with the electronic systems now contributing 90% of automotive innovations and new features, from emission levels to active-passive safety systems and entertainment-connectivity features.

The automotive today uses electronic controls for several applications including braking, airbag deployment, power steering, transceivers and infotainment. In the last decade, the improvements in IC design for safety to the passenger and driver in the automotive had led to a huge development of integrated smart power drivers (SPD). These drivers not only switch and control electric power but must be rugged to withstand fault conditions and continue functioning normally upon removal of the fault. Surveys did show that the automotive sector significantly outperformed the overall market average for semiconductors in 2014 [2]. It is one of the largest end markets for the power semiconductor applications ($7B). Automobiles produced today have a longer life expectancy than at any time in the past. This is due to the improvements in materials and design on one side. On the other side, the strongest contributor is the increase in electronic equipment and systems that

© Springer International Publishing AG 2018
S.N. Easwaran, *Current Sensing Techniques and Biasing Methods for Smart Power Drivers*, https://doi.org/10.1007/978-3-319-71982-5_1

Fig. 1.1 Worldwide production of electronic equipment dedicated to automotive (in billion euros) [1]

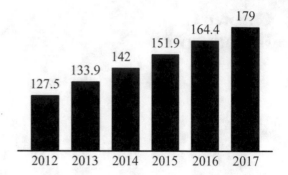

have replaced mechanical devices. Without proper electrical protection, however, these electronic systems in automobiles can fail without warning thereby putting the driver and passengers under huge risk. In 2011, ISO262 development was introduced to design ICs based on the safety flow in order to further enhance the safety in automotive and is steadily expanding to industrial applications as well.

The higher growth rate of automotive market is attributed to the following factors:

- The automotive industry is entering a new age of electronics with new safety and connectivity system solutions to boost the penetration rate of electronics in the automotive.
- There is a relentless and strong demand for improving energy saving, pollution control and road safety which can be satisfied only by electronic solutions.
- The new vehicle generations in general have a rapidly growing electronic content, as the most recent security and comfort equipment is becoming standard in the mid-range and lower-end vehicles.
- The development of hybrid and electric vehicles with a higher electronic content will have an increasing impact on the global automotive electronics market during the forecast period; the impact of this development will continue to increase between 2017 and 2020 and should become prevalent beyond 2020.
- Finally the digital automotive is dominated by connectivity and mobility.

The connected automotive is expected to provide the better solutions to road safety and intelligent transport while enhancing passenger's information and entertainment capabilities.

1.1   Automotive Electronic System

Electronic technology is now applied to a diverse spectrum of automobile operations [3], helping to improve driving performance, fuel efficiency, emissions purification, comfort and pleasure. Figure 1.2 shows the electronic content in an automobile. In an automobile or car, there are several types of automotive

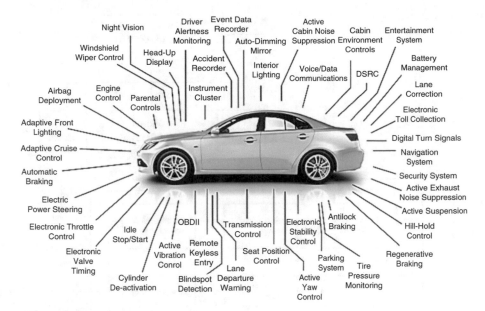

Fig. 1.2 Electronic content in automotive today [3]

electronic control units like engine control unit (ECU), airbag control unit (ACU), inverter safety unit (ISU), advanced driver assistance system electronic control unit (ADAS-ECU) and safety control module (SCM). The electronic control units in automotive are basically built up with the power supply unit, sensor unit, microcontroller and the driver. For example, a basic airbag control unit is built up with a power supply unit (PSU), sensors (crash, acceleration sensors) and microcontroller, and the squib drivers is shown in Fig. 1.3.

The PSU receives the battery voltage (V_{BAT}) as input. Several voltage rails are generated from the PSU which supply the sensor, microcontroller and the squib driver. The sensor senses any crash event and that information is passed on to the microcontrollers for processing in less than 50 ms time after which the software commands the squib driver in order to deploy the airbag. A detailed block diagram of the ACU [4] is shown in Fig. 1.4. The circuits like DC to DC converters, ADC, 5 V power supply, etc. that are needed to build this automotive electronic system are fundamentally not different to the electronic systems with which cell phones and laptops are made of. However, there are additional challenges that have to be considered for the automotive integrated circuit design. These are discussed in the following section.

High Power, High Junction Temperature Challenge

Most applications require the ambient temperature range of $-40\,°C$ to $-125\,°C$. In consumer applications, the ambient temperature range is typically $0\,°C$ to $-85\,°C$. When operating these electronic systems at high voltages and high currents, the power dissipation within the system increases. For instance, even under time-

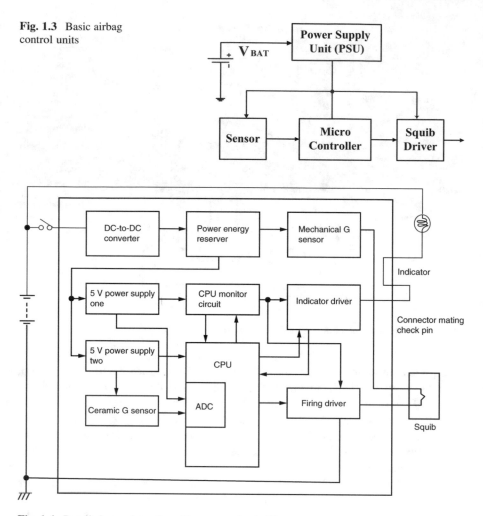

Fig. 1.3 Basic airbag control units

Fig. 1.4 Detailed overview of an airbag control unit [4]

limited conditions, a 35 V supply driving a current of 2 A for 0.7 ms will dissipate 70 W inside the electronic system for 0.7 ms. This will increase the junction temperatures of the transistors within the system [5]. There are several applications like motor drivers, power converters where the level can be 100 V or even higher. These electronic systems should be capable of operating as desired at these temperatures without any destruction.

Multiple Supply Voltage (MSV) Designs and Power Fault Challenge
Modern-day electronic systems require independent voltage supplies. For example, state-of-the-art digital camera usually needs about 15 different supply source voltages. The battery voltage is converted to all required voltage levels with efficient voltage regulators. The same applies to the automotive electronic circuitry

as it needs to be interfaced to the battery and also to the microcontroller. When these different voltage regulator outputs are used to drive different modules of an electronic system, interfacing becomes critical. For example, level shifting signals from a low voltage (LV) to a medium voltage (MV) or to a high voltage (HV) has to be handled with care in order to avoid inadvertent turn *on* or turn *off* of a system. Inadvertent turn *on* is extremely dangerous where the activation of a driver stage unintendedly happens. For example, in the ACU, this is a situation where the airbag can misfire.

Load Fault Challenge
The loads to the electronic system can be at fault. The probability is high since these load points can get shorted to battery or shorted to ground. Hence the overall electronic system should be fault tolerant. This fault-tolerant condition should be valid for the device in both the powered and unpowered state of the ASIC. These load points are inductive. The free-wheeling currents have to be provided by the electronic system during turn *off* conditions without any external Schottky diodes.

Diagnostics Challenge
The diagnostic section of the electronic system can fail without any warning to the driver or passenger in an automobile. Hence the electronic systems should be capable of operating as desired to provide meaningful diagnostic information under the power fault scenario, high temperature conditions and the load fault conditions.

    In summary, an automotive electronic system should be designed to achieve its main functionality along with being fault tolerant and is able to control inadvertent current. Each electronic system should have its built-in diagnostics in order to provide a warning without fail under these fault-tolerant modes of operation. The thesis is based on the research work performed for the development of integrated circuits for driving the squibs for airbag applications demanding fault tolerance and inadvertent current control. This thesis focusses on the current regulation circuits in airbag squib drivers with fault-tolerant topologies and inadvertent current control. Nevertheless these fault conditions are valid for automotive ICs used for several applications like braking, power steering, power trains, etc.

1.2   Squib Driver Function

A squib driver consists of two smart power switches through which a time-limited constant current is driven into the squib (energy less than 8 mJ [6]) in order to inflate the airbag when a crash is detected. Two switches are needed for redundancy and safety so that a single point failure will not fire the squib [6]. The configuration is that one power switch is configured as a high-side FET (HS_FET) and the second switch is configured as LS_FET (low-side FET).

    The drain of the HS_FETs, i.e. the VZx pins, is connected to the energy reserve directly or through a third switch which is activated prior to deployment. This is

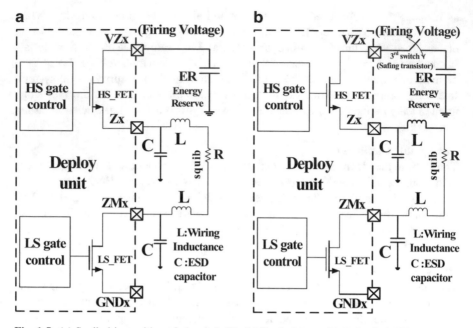

Fig. 1.5 (**a**) Squib driver without 3rd switch [8]. (**b**) Squib driver with 3rd switch [8]

shown in Fig. 1.5a, b. The current through the HS_FET is regulated in order to ignite the squibs to inflate and deploy the airbag.

State-of-the-art driver IC in the ACU has integrated sensors and squib drivers. Some driver ICs in the ACUs integrate the PSU in addition to the sensors and the drivers. These are called as the SBC (System Basis Chip). The PSU output supplies the on-chip sensors and drivers in addition to the airbag squib driver extension ASICs (application-specific integrated circuits). This is shown in Fig. 1.6.

1.3 Squib Driver True Deployment and Inadvertent Deployment Conditions

To deploy the squibs, the current through the HS_FET is regulated between 1.75 A and 2.14 A for 0.5 or 0.7 ms and between 1.2 A and 1.47 A for 2 ms [7, 8]. The LS_FET is operated in R_{ds_on} mode but includes a current limitation (higher than HS_FET) in order to be protected during a fault condition where the drain of the LS_FET can get shorted to battery. The resistance of the squib can vary between 1 Ω and 8 Ω during deployment event. This is the condition for deployment of the airbag for a true firing event. In other words, a time-limited energy of around 8 mJ is needed in order to deploy the squib. This is shown in Fig. 1.7.

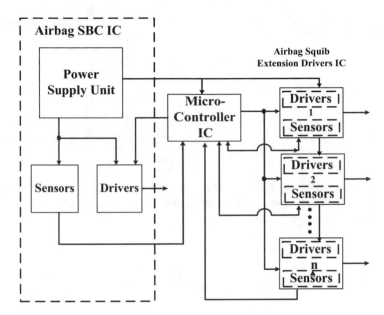

Fig. 1.6   Squib driver SBC and extension ASICs

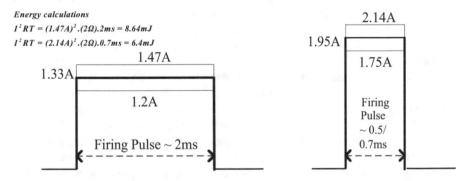

Energy calculations
$I^2RT = (1.47A)^2.(2\Omega).2ms = 8.64mJ$
$I^2RT = (2.14A)^2.(2\Omega).0.7ms = 6.4mJ$

Fig. 1.7   Current regulation targets for deploying airbag

On the other hand, to avoid the misfire of the airbag or in other words called as the inadvertent deployment, the energy in the squib should be limited to ∼175 µJ [9] in powered, partly powered and unpowered state of the driver when a short circuit to battery occurs on the LS_FET or a short circuit to ground occurs on the HS_FET. The equivalent behaviour is to have the peak current limited to 5 A for 4 µs as per AK-LV16 standard [10]. This is shown in Fig. 1.8.

In addition, the quiescent current of the driver IC should be within the targeted specification under any power-up condition. The target is 5 mA maximum. The diagnostic module which is a critical part of the driver IC needs a precise reference voltage for diagnostics that is independent of the fault conditions and needs a

Fig. 1.8 Current limitation
targets for not deploying
airbag

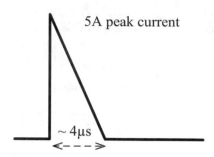

precise clamping level at its output such that the signal processing stages like the
ADC does not face any overvoltage conditions that can lead to inadvertent current
flow or destruction of the ASIC.

1.4   Current Sensing and Regulation

Current sensing is one of the critical building blocks in power electronic circuits.
For switching converters, drivers, etc., there is always a need for overcurrent
detection or current limitation in order to protect the system against overcurrent
events. Current regulation circuit is essential to inflate airbags. Conventional
current sensing techniques employ a sense resistor for sensing the current in order
to regulate or limit the current. Since the squib resistance itself varies dynamically
during the deployment event, a current sense approach using the squib resistance
itself is not reliable. Hence a resistor independent of the squib has to be used.
Employing an external resistor in series with the squib is not an option as it will
demand an additional pin to the driver IC for the current sensing scheme. This is not
area- and cost-efficient when it comes to a device involving multichannel drivers
[11]. One of the earlier solutions employs an integrated poly resistor at the source/
emitter of the power switch in order to sense the current flow. The voltage drop
across the resistor activates an NPN transistor such that the current through the
power switch is limited to 0.7 V/R [12]. However due to the flyback effect in
the presence of the inductive load leading to substrate currents, the current sense
resistors are connected to the drain of the powerFET to sense the current. This
approach is limited to lower currents (100–300 mA) as it impacts the die area.

Some topologies employ current sensing by utilizing the bondwire resistance.
This concept can be used for very high currents (>10–15 A). It is important to note
that the resistor should be wide enough to conduct the current. So it is a trade-off
between area and performance. For currents in the order of 2 A, this resistance is too
low and becomes critical as the detected signal is overwhelmed by noise and
offsets.

To overcome the disadvantages of poly resistor-based current sensing, R_{on}- and
sense FET-based current sensing schemes were introduced. The FET R_{on} [13]

technique estimates the current from the drain – source voltage of the MOS and hinges on the resistance variation of the switch. This resistance variation is significant (50–200% variation) with temperature, process and supply and hence not accurate. The senseFET-based current sensing is based upon sensing a portion of the main current using a current mirror, whose mirroring is on the order of 300 or even higher. The accuracy is dependent on the matching of the mirrors and is less accurate with regard to the sense resistor approach but much better than the R_{on}-based sensing scheme. For the squib driver application, the current regulation on the HS_FET needs to be accurate ($\pm 10\%$). Currents lower than 1.75 A or 1.2 A will not be sufficient to inflate the airbag. Currents higher than 2.25 A for 1.75 A condition or 1.5 A for 1.2 A condition will deplete the energy reserve faster and is not preferred. A current limit on the LS_FET is necessary for protection against short to battery faults and should be higher than the current limit of the HS_FET. The accuracy level can be $\pm 15\%$. The LS_FET should be able to thermally sustain without any destruction.

The current sensing schemes should be able to deliver stable currents despite a wide variation of the inductive and capacitive loads. This drives the choice behind the topologies for the current sensing scheme in the HS_FET and LS_FET. However this current limitation scheme is obviously effective only when the device is in the powered state, i.e. when all the supplies, gate driver supply and the bias currents are available. In situations where the device is partly powered or unpowered, energy limitation is mandatory so that the airbag is not inadvertently deployed. Hence a current regulation loop on both HS_FET and LS_FET needs to be implemented along with the energy limitation to avoid inadvertent deployment during fault conditions in the partly powered or unpowered state of the IC. This is shown in Fig. 1.9a, b independently for the HS_FET and LS_FET. The combination of current regulation scheme in the powered state and the current limiting scheme in the unpowered state is the driving force behind the proposed squib driver architecture. Table 1.1 provides an overview of different current sensing methods.

1.5   Biasing Scheme

Having discussed the goals for combination of current regulation scheme in the powered state and the current limiting scheme in the unpowered state, the partly powered situation in the squib driver IC is a grey area of operation since the biasing conditions of several transistors, in the MSV domain, could be undefined. The designs have to be developed to withstand faults like missing power supply in a multiple supply domain ASIC that can lead to high currents due to floating nodes. These circuits have to be fault tolerant irrespective of different sequence of power supplies and guaranteed not to misfire airbags when some supplies are either missing.

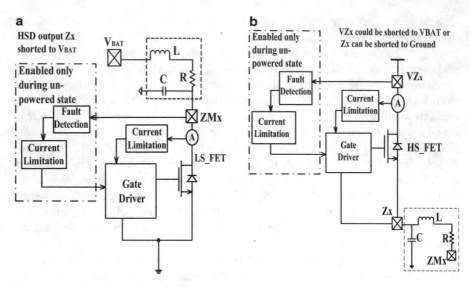

Fig. 1.9 (**a**) Low-side driver. (**b**) High-side driver

Table 1.1 Comparison of current sensing methods

Method	Description	Advantages	Disadvantages
Sense resistor source side	Poly resistor in series with the switch	Accurate. Regulation loop stability is easier as the architecture is similar to a source follower type	Sensitive to substrate currents. Limited to less than 300 mA
Sense resistor drain side	Poly resistor in series with the switch	Accurate and insensitive to substrate currents	Stabilizing the loop is not straightforward as the power stage is a gain stage
Power switch R_{on}	R_{ds_on} of the switch	Higher-current limits can be achieved	Inaccurate
SenseFET	Scaled version of powerFET for current sensing	Higher current limits can be achieved	Threshold voltage mismatch between powerFET and senseFET

1.6   Voltage Reference and Voltage Clamping

The diagnostic module which is a critical part of the driver IC needs a precise reference voltage for diagnostic functions like the squib resistance measurement [14]. This reference V_{REF} is a current limited voltage source (CLVS) and should be fault tolerant. Fault tolerant refers to protecting the circuit against short-to-ground or short-to-battery faults on the squib pins Zx or ZMx.

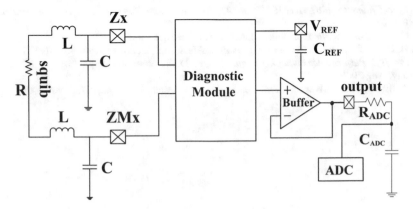

Fig. 1.10 Diagnostic module with output buffer

A precise clamping level (less than 5.5 V) at the diagnostic output is needed such that the signal processing stages like the ADC will not be subjected to overvoltage conditions. These overvoltage conditions can lead to inadvertent current flow or destruction of the ASIC. The block diagram is shown in Fig. 1.10.

References

1. D. Coulon, Whatever the future of the automotive industry, electronics is the key, *TTI/ Market Eye*, 9 Sept 2014
2. J. Liao, Automotive industry now the third largest end market for power semiconductors [Online, IHS Press release] (2015). Available FTP: https://technology.ihs.com/548523/auto motive-industry-now-the-third-largest-end-market-for-power-semiconductors-ihs-says
3. CVEL (Clemson University of Vehicular Electronics Laboratory), Automotive electronics [Online]. Available FTP: http://www.cvel.clemson.edu/auto/systems/auto-systems.html
4. M. Hitotsuya et al., Electronic control unit for a single-point sensing airbag. Fujitsu Ten Tech. J. **7**, 13–18 (1995)
5. P.L. Hower, Safe operating area – A new frontier in LDMOS design, in *Proceedings 14th International Symposium on Power Semiconductor Devices and ICs*, Santa FE, New Mexico, June 2002, pp. 1–8, doi: https://doi.org/10.1109/ISPSD.2002.1016159
6. R.S. Wrathall, A study of AC and switch mode coupling of currents to airbag squib ignitors, in *Proceedings Workshop on Power Electronics in Transportation, IEEE*, 1996, pp. 111–116, doi: https://doi.org/10.1109/PET.1996.565918
7. *Airbag Combined Power-Supply and Firing Circuit*. TLE6710Q Datasheet, Infineon Technologies, Version B, 4 May 2001
8. *Quad Channel Driver for Airbag Deployment*. TPIC71004-Q1, Datasheet, Texas Instruments, SLVSAT2, Feb 2011
9. V. Colarossi, New development in Autoliv squib driver validation process, in *12th International Symposium and Exhibition on Sophisticated Car Occupant Safety Systems*, Karlsruhe, Germany, Dec 2014
10. *AK-LV-16*, Electrical Igniters for Pyrotechnical Systems, Requirements and Test Conditions. Specification of the Automotive Industry

11. *Four Channel Squib Driver IC. MC33797, Technical data, Rev 6.0, Freescale Semiconductor*, Feb 2014
12. *Current Loop Application Note*, CLAN1495, B&B Electronics, 1995
13. H. Pooya Forghani-Zayed, *An Integrated, Lossless, and Accurate Current-Sensing Technique for High-Performance Switching Regulators*, Ph.D. dissertation. Georgia Institute of Technology, Aug 2006
14. A. Sibrai, Squib driver for airbag application, U.S. Patent US 7,142,407, 8 Nov 2006

Chapter 2
State-of-the-Art Current Sensing, Biasing Schemes and Voltage References

Current regulation circuit is essential for the squib drivers to inflate airbags. Energy less than 8 mJ should be provided to the FETs to deploy the airbag. These are applicable when the driver is powered and the ACU microprocessor commands to deploy the squibs. The HS_FET is needed to provide a precise regulated current of 1.95 A (1.75 A $< I_{\text{deploy\_HI}} <$ 2.14 A) for 0.5/0.7 ms and 1.35 A (1.2 A $< I_{\text{deploy\_LO}} <$ 1.47 A) for 2 ms. The LS_FET has a current limitation of 3 A ±15%. Based on which setting is used, i.e. either $I_{\text{deploy\_HI}}$ or $I_{\text{deploy\_LO}}$, the duration of the current limit on the LS_FET is set for a short-to-battery condition on the drain of the LS_FET. In the unpowered or partly powered (missing supplies) scenarios, the airbag misfire should not occur, and hence additional protection is needed. However, as will be discussed in this chapter, the realization of a current regulation loop and a current limitation in unpowered state, biasing scheme, diagnostic circuits exploiting voltage and current selector circuits is not currently available and is therefore the objective of the present research.

2.1 State-of-the-Art Current Sensing Schemes HS_FET

Conventional current sensing techniques for airbag drivers employed current sensing on the source of the FETs. This is shown in Fig. 2.1 [1]. Another current sensing architecture [2] for airbag squib drivers requires a SENSE pin in order to provide the current limitation. However the resistor on the source side of the HS_FET is sensitive to negative voltages during turn *off*, and the SENSE pin-based approach is expensive for multichannel drivers. The disadvantages of the sense resistor on the source side were overcome by sensing the current on the drain by using the sense resistor Scheme [3]. Some squib driver ICs [4] use a senseFET-based current regulation. This is shown in Fig. 2.2. The OTA1 ensures that the source of the powerFET and senseFET are equal for accurate current sensing. This results in a

© Springer International Publishing AG 2018
S.N. Easwaran, *Current Sensing Techniques and Biasing Methods for Smart Power Drivers*, https://doi.org/10.1007/978-3-319-71982-5_2

Fig. 2.1 Current sensing
scheme using sense resistor
at the source of the
powerFET [1]

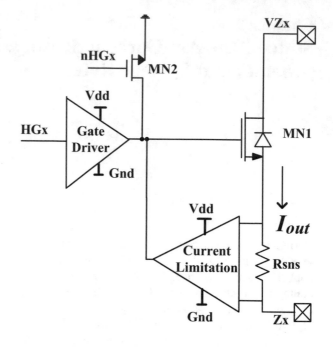

Fig. 2.2 Current sensing
scheme using senseFET [4]

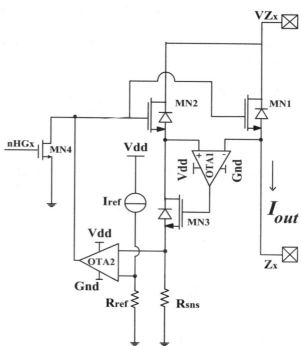

scaled version of the output current I_{out} through the senseFET. This current creates a drop on the resistor R_{sns} and is regulated to the voltage $V_{ref} = I_{ref}.R_{ref}$ by OTA2 [5].

In order to limit the current during the unpowered state so that inadvertent activation of the powerFET is prevented, an additional discharge path at the gate of the HS_FET is required. Unpowered state refers to a condition when Vdd is 0 V or relatively low such that pull-down or pull-up transistors are not turned *on* to ensure the HS_FET is turned *off* completely. Mechanism to ensure that the powerFET does not turn *on* when the drain voltage of 12 V is applied suddenly on the VZx pin or when a transient on the drain of the HS_FET occurs is important. These transients can be in the order of 50 V/μs or higher. So it is critical to ensure that the V_{gs} of the HS_FET is less than its threshold voltage V_{th} despite a high voltage on the powerFET drain will require additional circuitry that can be implemented by using a passive element like resistors or with a combination of active elements like transistors.

2.2   State-of-the-Art Current Sensing Schemes LS_FET

A similar concept is applied for the LS current sensing. The reduced circuitry is shown in Fig. 2.3.

Since LS current sensing need not be as precise as the HS current sensing, OTA1 can also be removed [3]. When current through MN1 increases, the current sensed through MN2 also increases. This increases the voltage drop across the sense

Fig. 2.3 Current sensing scheme using senseFET without OTA1 [4]

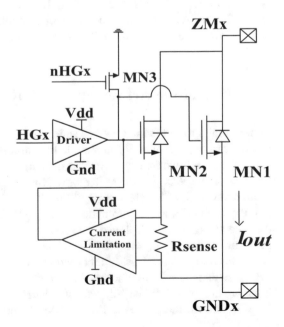

Fig. 2.4 Current sensing
scheme using senseFET
without OTA1 derived from
Fig. 2.2 [4]

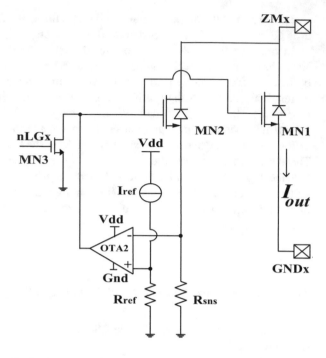

resistor R_{sense}. The current limitation circuit decreases the gate voltage of MN2 and
MN1. The decrease in gate voltage decreases the current through MN1 such that the
current does not exceed the targeted value. A modified current sensing Scheme [4]
is shown in Fig. 2.4. This is used for the LS current sensing scheme as the limit is
wider. The OTA1 is not needed if a precise regulation is not needed and the R_{sns}
being a low value that will make the drop across few tens of mV higher than GNDx.

2.3   State-of-the-Art Energy Limitation

Energy limitation in the unpowered state is becoming a critical requirement in the
recent years to prevent airbag misfire. In the unpowered state, there is no supply and
hence the pull-down transistor MN1 is not active anymore. If a fast short to battery
happens on the VZx pin, the miller capacitance C_{gd} will be charged due to the fast
dV/dt and will rise the gate voltage of the HS_FET. This will trigger a peak current
into the squib prior to the current decreasing to 0. If this energy is higher than 8 mJ,
then the airbag will be deployed. The energy limitation circuit is to ensure that the
peak current through the HS_FET (which is also the peak current through the squib)
is restricted to 5 A for 4 µs. In other words, it is needed to keep the energy below
170 µJ so that the misfire of airbag or inadvertent deployment does not occur. This
is shown in Fig. 2.5. When Vdd and HGx levels are 0 V and VZx has a fast step of
35 V in 300 ns, the goal is to limit the MN1 current I_{out} to 5 A for 4 µs. This is

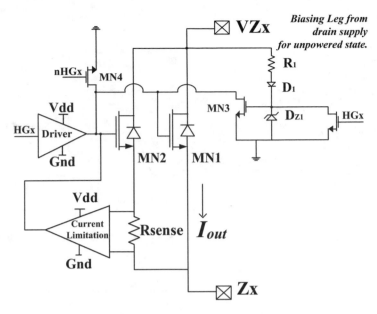

Fig. 2.5 Energy limitation circuit [4]

achieved by forming a bias network [4] from VZx through R_1, D_1 and D_{Z1} where D_{Z1} is a 6 V or 12 V Zener diode. This keeps MN3 active to pull down the gate of the HS_FET. During the powered state, the HGx signal when activated will keep MN2 off and will not interfere with the main current regulation loop.

2.4 State-of-the-Art Biasing Scheme in Multiple Supply Voltage Domains

Automotive designs involve multiple supply voltage domains involving low-voltage, medium-voltage and high-voltage rails. However these supplies do not ramp at the same time creating undefined or tri-stated conditions that can inadvertently turn *on* the drivers. This will result in very high currents, and so there is a need for current control. State-of-the-art solutions [6] employ switching of the bias currents by using the Bandgap_Ready signal. The second bias block receives its input bias current based on the Bandgap_Ready signal. If this signal is 0, the Bias 1 serves as the input bias current to the second bias block; else the Bias 2 is used. This is shown in Fig. 2.6.

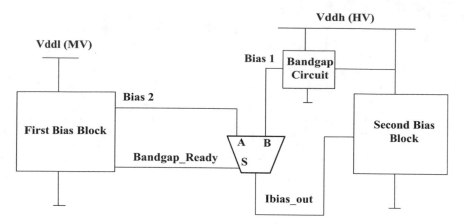

Fig. 2.6 Biasing schemes in multiple supply voltage designs with multiplexer approach [6]

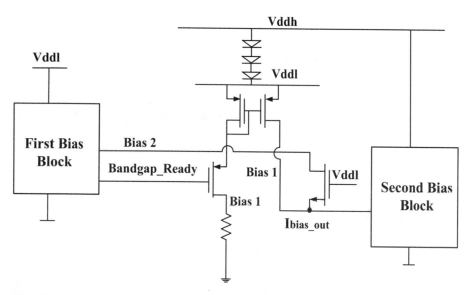

Fig. 2.7 Biasing scheme in multiple supply voltage designs with current comparator approach [7]

Additional biasing schemes reported at [7, 8] do follow a similar approach. Instead of using the Bandgap_Ready signal, the scheme utilizes a comparator-based decision signal to switch between the bias currents. The comparator compares the Bias 1 and Bias 2 currents and, based on the decision, decides if the start-up bias current or the main bias current has to be provided. This is shown in Fig. 2.7.

2.5 State-of-the-Art Biasing Scheme for Diagnostics

The current limited voltage sources and output voltage limitation are needed for the diagnostic module. Implementation of a voltage reference with a current limitation is shown in [9]. The voltage on the external resistor R_{EXT} is regulated to the reference voltage V_R to generate a constant reference voltage. When a fault like the short to ground occurs on the external resistor, the voltage is pulled low with the current being limited by the current mirror at the top. This is shown in Fig. 2.8.

For airbag applications, a similar architecture is presented in [10]. The CLVS (current limited voltage source) shown in Fig. 2.9 is needed to measure the resistance of the squib. However the regulation has to be performed at the drain of the FETs in order to achieve the common mode regulation voltage in order to set the common mode level for the Zx and ZMx pins for the squib resistance measurement. It should be ensured that the impedance from Zx to GND should be lower than the impedance from Vdd to Zx. This way I_p is the dominant current and the voltage on Zx is regulated to V_R level. It is typically around 8 V. V_{R2} is an internal reference that is typically generated from on-chip bandgap reference circuit. This referencV_{R2} is used to bias the PNP transistor. This will ensure that a large proportion of the current I_p flows into the PNP transistor since the PNP transistors have more current carrying capability than the diode path.

Fig. 2.8 Current limited
voltage source [9]

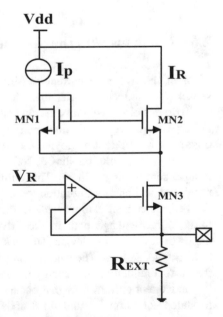

Fig. 2.9 Current limited
voltage source for squib
resistance measurement
[10]

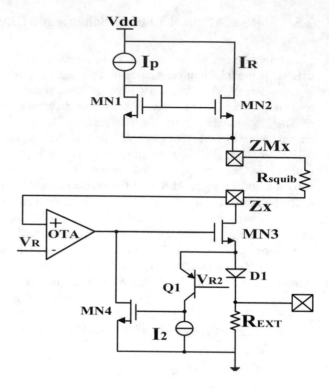

2.6   Problems with the State-of-the-Art Techniques

Most of the available current sensing circuits use either a sense resistor or a
senseFET-based current scheme. However not all available circuits for airbag
squib drivers effectively mix the current sensing scheme in powered state to deploy
airbag with energy limitation circuitry in the unpowered state to avoid airbag
misfire or inadvertent deployment during fault conditions like the dynamic short
to battery. The solution that effectively handles this is through the use of an
independent biasing path [4]. This eliminates the interference between the current
regulation and energy limitation loop. The disadvantage with this approach is that
the leakage current is higher due to the permanent current path. This current will
mask a true leakage path in the driver circuits connected to the VZx pin in
unpowered state of the device. The leakage current on this pin is typically specified
to be ±5 μA at 40 V. The solution proposed in this dissertation effectively handles
the current regulation and energy limitation with low leakage current.

Inadvertent current flow that occurs in the smart power drivers is triggered by
tri-stated conditions or undefined nodes during the fault conditions like missing
supplies or the absence of some of the supplies in a multiple supply voltage design.

Existing solutions use control signal-driven decision regarding the biasing scheme. This suffers the disadvantage of expensive filtering to eliminate noise or suffers performance due to glitch in the control signal.

A control signal-independent biasing scheme is proposed in this dissertation by exploiting the current selector circuits. The current selector solution discussed here helps overcoming the disadvantages of the state of the art. A precise output voltage clamping method that is robust for input swings from 100 mV to 30 V or higher at very high on-chip junction temperature is discussed.

In this book, the implementation of a monolithic HS_FET and LS_FET current regulation scheme combined with fault-tolerant and inadvertent current control is discussed. This circuit implementation is needed for driving the airbag squib drivers but can be applied to several automotive and other safety applications. It starts by investigating the current sensing schemes individually for the HS_FET and LS_FET based on the accuracy needed and the stability requirements of the current regulation loop for a wide range of inductive loads. Once the current sensing scheme has been finalized, the current limitation circuitry for avoiding inadvertent current flow through HS_FET and LS_FET under fault conditions is explored and added to the gate control of the HS_FET and LS_FET.

It is important to ensure that both loops do not interfere with each other, i.e. the current regulation in powered state to deploy airbag should not be disturbed by the current limitation circuit for unpowered state. The effect is that the interference could cause the airbag not getting deployed when it is needed and getting deployed when it is not needed. This is a safety violation and can result in driver and passenger injuries and is a life threat. The biasing scheme with a novel topology to ensure fault-tolerant behaviour is discussed in detail in order to ensure that the operating conditions are well defined. In multiple supply voltage (MSV) designs where different power domains are available at different times, it is critical to eliminate undefined tri-stated outputs or floating nodes that can trigger inadvertent currents.

The diagnostic circuitry for the squibs is discussed in detail where the necessity of a current limited voltage source (CLVS) is discussed. This is extensively needed for the squib resistance measurement (SRM) as this is diagnosed prior to deployment. For instance, if there is no passenger in the car other than the driver, the seat belt on the passenger side is not worn indicating the resistance is high in the order of 1 KΩ instead of being less than 8 Ω. Hence there is no necessity to deploy the passenger side airbag during a crash event. Deployment of the driver side airbag is sufficient to protect the driver.

The output of this diagnostic module is processed through an ADC. Even though the input of the diagnostic module can vary from 100 mV to 30 V, the output of the diagnostic module or the ADC input has to be clamped to 5.5 V to avoid overvoltage stress that can overstress the 5 V ADC and trigger high currents. A precise voltage clamping method is presented to address this issue. All these circuits were simulated and validated on a four-channel squib driver in a 40 V, 0.35 μm Texas Instruments (TI) BiCMOS process. BiCMOS process supports LV, MV rated NPN, PNP transistors, LV, HV CMOS transistors and LDMOS transistors. It includes

poly resistors with low-to-high sheet resistances, poly-to-N+ diffusion capacitors and HV metal-to-metal capacitors. The following section describes the specifications of an airbag squib driver IC.

2.7   Specification

Table 2.1 details the specification of the airbag squib driver ICs. It specifies the deployment current during the deployment mode of operation to deploy the airbag, the surge current limitation.

2.8   Chapters Organization

Here in this book, a current regulated squib driver circuit with fault-tolerant and inadvertent current control is presented. Several design challenges encountered while designing the driver, its biasing concept and diagnostic circuitry are outlined.

In Chap. 3, the principles related to current regulation and current limitation circuits are reviewed. The fundamental issue in level shifting schemes in multiple supply voltage designs is discussed.

Table 2.1 Specification of squib drivers

Spec parameter	Target	Comments
HS_FET current regulation	2 A ±10% and 1.35 A ±10%	Need both settings for airbag deployment. Inductance range 1 μH to 3 mH.
LS_FET current regulation	3 A ± 15%	Protection for LS in case of short circuit to battery (SCB). 1 μH to 3 mH
HS_FET surge current limitation	5 A,4 μs	To prevent inadvertent deployment
LS_FET surge current limitation	5 A,4 μs	To prevent inadvertent deployment
Quiescent current of the driver	5 mA	No diagnostics are enabled
Diagnostic reference (CLVS)	8.4 V	Cap free 100 pF to 220 nF
Output voltage clamping	5.4 V	Input can range anywhere between 100 mV and 30 V
Deployment voltage	18–33 V	Typical deployment voltage is 33 V (energy reserve capacitor output)
Leakage current on drain of HS_FET	< 1 μA	Range 0–35 V
Leakage current on drain of LS_FET	< 1 μA	Range 0–35 V

In Chap. 4, the choice of the current sensing scheme by using a metal sense resistor for the high-side driver is presented along with stability analysis. The stability for a wide range of inductive loads and the internal free-wheeling path is presented. The inadvertent or surge current control circuit that is needed to control inadvertent deployment is discussed and the proof of independency of the two circuits is outlined.

In Chap. 5, the current sensing scheme for the low-side driver by using a senseFET is also presented. The choice of the architecture to support wide inductive loads is presented along with the small-signal analysis. The topologies used for the V_{ds} equalizer and the regulation loop are presented. There are no dedicated compensation capacitors, and the loop is stabilized by utilizing the intrinsic powerFET capacitance. A novel topology for surge current control is presented.

In Chap. 6, a novel biasing scheme involving a maximum current selector and a fault mode supply generation circuit is presented to avoid any undefined bias condition for the HS and LS current regulation loops. This chapter stresses the need for the improvements to the state of the art and introduces the maximum current selectors. It also explains why these circuits could be exploited for ensuring the well-defined bias conditions for the electronic circuits and preventing inadvertent currents. Structures for the proof of concept are simulated and measured. The diagnostic module with a fault-tolerant current limited voltage source and an improved output level clamping technique despite a huge input voltage swing is presented. Proposed design techniques have been implemented in a 0.35 μm BiCMOS process.

Lastly Chap. 7 provides conclusions and describes the future work.

References

1. *Dual Airbag Deployment ASIC*. CS2082-D, Datasheet, Rev 7.0, ON Semiconductor, July 2001
2. *Four Channel Squib Driver IC*. MC33797, Technical data, Rev 6.0, Freescale Semiconductor, Feb 2014
3. *User Configurable Airbag IC*. L9678, L9687-S, Datasheet, DocID025869, Rev 3.0, STMicroelectronics, May 2014
4. *4 Channel Squib Driver*. E981.17, Datasheet, QM-No. 25DS0073E.00, Elmos, Aug 2011
5. *2-Ch Squib Driver For Airbags*. TPD2004F, Datasheet, Toshiba, Oct 2006
6. C.Y. Chen, System and method for breakdown protection in start-up sequence with multiple power domains, U.S. Patent US 7,391,595B2, 24 June, 2008
7. Y. Date et al., Current driver, U.S. Patent US7,466,166B2, 16 Dec 2008
8. B. Sharma et al., Avoiding excessive cross-terminal voltages of low voltage transistors due to undesirable supply-sequencing in environments with higher supply voltages, U.S. Patent US 7,176,749B2, 13 Feb 2007
9. F. Pulvirenti, Programmierbare Schaltung mit Strombegrenzung fur Leistungsstellantriebe, German Patent DE 69613956T2, 4 Apr 2002
10. *Airbag Combined Power-Supply and Firing Circuit*. TLE6710Q Datasheet, Infineon Technologies, Version B, 4 May 2001

Chapter 3
Current Sensing Principles with Biasing Scheme

State-of-the-art electronic circuits heavily utilize the information related to the amount of current being delivered to the load. In safety-critical automotive applications like the airbag squib drivers, it is critical to utilize this information in order to regulate the current that is delivered to the squib. To know about the current that is delivered, a current sensing technique is needed. A sense resistor is typically used to sense the amount of current being delivered to the load, or a senseFET is used in order to sense the amount of current being delivered to the load. In this chapter, the fundamentals of current sensing and how this sensing concept is used to limit the current are being reviewed.

3.1 Current Sensing Circuit with Sense Resistor

The conventional current sensing circuits sense the amount of current being provided to the load and use the sense resistor concept. This is shown in Fig. 3.1. The resistor R_{sense} is placed in series with the load. The V_{out} voltage that carries the information about the load current is derived as per the equation below. This type is called as the high-side current sensing [1] since the sense resistor is connected between the load and the supply, V_{BAT}.

The load current, I_{load}, drops across the resistor R_{sense} to create a voltage drop V_{load}. Since the inputs of the AMP are at virtual ground, V_{out} can be derived as shown below and is determined by the ratio of R_{out} and R_2.

$$V_{BAT} = (I_{load}R_{sense}) + V_{load} \qquad (3.1)$$

Since AMP inputs are at virtual ground, the positive and negative terminals of the AMP are at V_{load}.

© Springer International Publishing AG 2018
S.N. Easwaran, *Current Sensing Techniques and Biasing Methods for Smart Power Drivers*, https://doi.org/10.1007/978-3-319-71982-5_3

Fig. 3.1 Conventional
current sensing circuits [1]

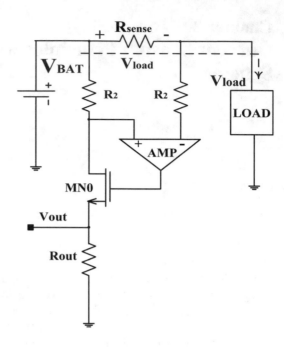

$$V_{\text{out}} = \left(\frac{V_{\text{BAT}} - V_{\text{load}}}{R_2} \right) R_{\text{out}}$$

$$\Rightarrow V_{\text{out}} = \left(\frac{V_{\text{BAT}} - [V_{\text{BAT}} - (I_{\text{load}}R_{\text{sense}})]}{R_2} \right) R_{\text{out}}$$

$$V_{\text{out}} = I_{\text{load}}R_{\text{sense}} \left(\frac{R_{\text{out}}}{R_2} \right) \tag{3.2}$$

A similar concept for low-side current sensing is shown below. In the low-side current sensing scheme, the sense resistor is placed between the load and ground. The voltage drop across the resistor R_{sense} is sensed through the AMP, and the resistor networks R_1 and R_2 are connected between the output of the AMP V_{out} and V_{sense}. This is shown in Fig. 3.2.

$$\left(\frac{I_{\text{load}}R_{\text{sense}}}{R_1} \right) = \left(-\frac{V_{\text{out}}}{R_2} \right)$$

$$\Rightarrow V_{\text{out}} = - \left(\frac{I_{\text{load}}R_{\text{sense}}}{R_1} \right) R_2 \tag{3.3}$$

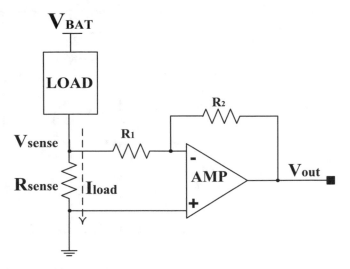

Fig. 3.2 Low-side current sensing [1]

3.2   Current Sensing Circuit with Sense Resistor and Current Limitation

The conventional current limiting circuits followed the principle of current sensing. In order to limit the current through $Q1$, a sense resistor at the source (for MOS-based powerFETs) or emitter of the bipolar-based powerFETs is used [2]. This is shown in Fig. 3.3.

The circuit has built-in current limiting so that the loop current cannot exceed I_{limit}. The resistor R_{g} senses the current flowing through $Q1$. It provides a source of bias current for $Q2$ so that if the loop tries to exceed I_{limit}, $Q2$ will shunt the $Q1$ base bias current so that $Q1$ cannot conduct more than I_{limit}. The I_{limit} value is set by the V_{BE} of $Q2$ and the resistor R_{g}. The disadvantage is the V_{BE} variation over temperature that will result in the current limit having a wide variation over temperature.

$$I_{\text{limit}} = \frac{0.7\text{V}}{R_{\text{g}}} \qquad (3.4)$$

3.3   Current Sensing Circuit with SenseFET and Current Limitation

The resistor R_{g} should be able to handle the current that is targeted to flow through it. For example, in Fig. 3.3, R_{g} has to be really wide to handle very large currents. In integrated circuits, a poly resistor area could be extremely wide to handle very high

Fig. 3.3 Conventional
current limiting circuits [2]

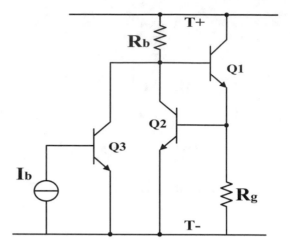

currents in the order of amperes. It is not area-efficient if the resistor is too large in area. This means that the current limitation achieved by using a poly resistor is in the lower current range of 200–300 mA. To overcome this problem, a senseFET-based current sensing is proposed where a smaller version of the FET is used in parallel to the powerFET to sense the current. This is shown in Fig. 3.4. The powerFET MN1 is N times bigger than the senseFET, MN2. This indicates that MN2 will ideally sense $(1/N)$ times the current through MN1. The OTA1 equalizes the gate-source voltage of the powerFET and the senseFET. This ensures that the current through MN3 will be (I_{out}/N) and drops across the resistor R_{sns}. The OTA2 regulates this voltage to the reference voltage generated by $I_{ref}.R_{ref}$.

$$\left(\frac{I_{out}}{N}\right) R_{sns} = I_{ref} R_{ref}$$

$$\Rightarrow I_{out} = \frac{I_{ref} R_{ref} N}{R_{sns}}$$

(3.5)

3.4   Biasing Schemes

For the current sensing and current limitation schemes to work, the OTAs should be biased in the appropriate way. These circuits should not get activated or deactivated unintentionally. In automotive terminology, it is called as the inadvertent activation or deactivation. These circuits will be present in the driver ASIC and is supplied by the power management ASIC as shown in Fig. 3.5. Hence there is a need for a robust biasing scheme.

The power management ASIC generates various output voltages from the battery (V_{BAT}) (~ 12 V for automotive applications) to supply the smart power driver ASICs. These different output voltages [3] are approximately 33 V, 6 V and

Fig. 3.4 SenseFET-based
current limiting circuit [5]

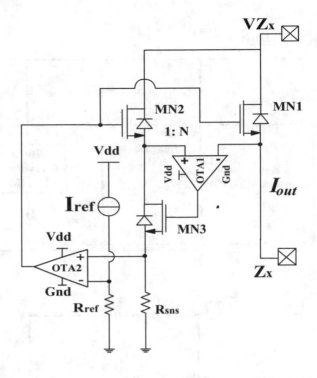

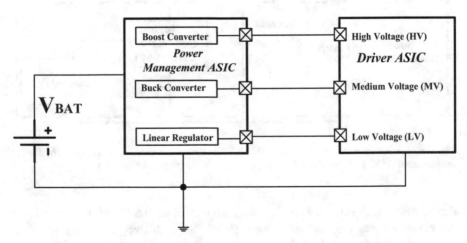

Fig. 3.5 Multiple supply voltage domains power supply concept

3.3 V. Circuits operated by multiple supplies are called as the "multiple supply voltage (MSV)" circuits. The 33 V rail is generated by the step-up or boost converter. 6 V level is generated by either the step-down or buck converter or low-dropout (LDO) regulator. The 33 V is the high-voltage (HV) supply for the

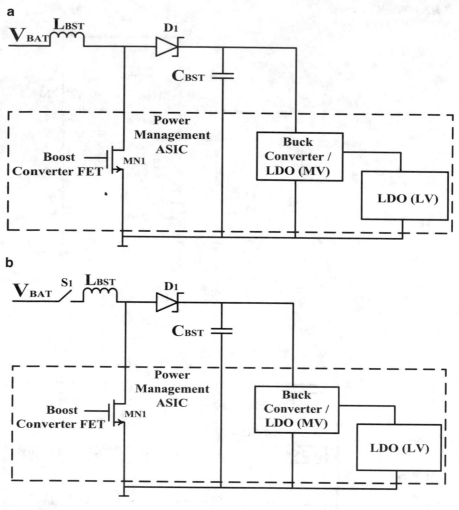

Fig. 3.6 (**a**) Power management ASIC with HV, MV and LV domain generators. (**b**) Power management ASIC with switch in the HV, MV and LV domain generators

driver ASIC. There are loads in the driver ASIC that need 5 V or 6 V supplies and are classified into the medium-voltage (MV) range. All applications do have a microcontroller (μC) that needs a low noise supply, and LDOs are used for these loads due to the high PSRR requirements. The driver ASIC needs 3.3 V or even lower voltage level to supply its digital core. This is classified as low voltage (LV). These different supply rails are not available at the same time for the driver ASIC. This is shown in Fig. 3.6. In other words, the delay between the availability of the supplies can be in several hundred microseconds (μs) to several tens of milliseconds (ms). The power sequencing will create tri-stated biasing conditions for circuits that

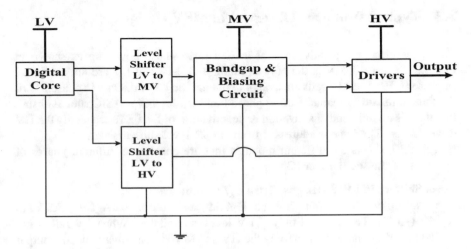

Fig. 3.7 Modules and supply domains in a driver ASIC

can activate unintentionally or inadvertently the power stages thereby leading to high currents. Such inadvertent turn *on* is very critical in airbag applications [4, 5] as a situation could occur where the driver gets turned *on* and deploys the airbag when there is no crash event that initiates the deployment. Similarly there could be unintentional turn *off* situations that can stop the activation of the driver.

Modules present in a power management ASIC are shown in Fig. 3.6a. The HV output is generated by a boost converter with its power stage MN1, L_{BST}, D_1 and C_{BST}. Voltage starts ramping from V_{BAT} (0.3 V) and is boosted to a typical output of 33 V. In some applications, there is a switch S_1 between the battery and L_{BST} as shown in Fig. 3.6b, to ensure that capacitor C_{BST} starts from 0 V. The output of the boost converter supplies the buck converter or LDO to generate the 5 V or 6 V level. The 5 V or 6 V level supplies the LDO to generate the LV level. The modules of the driver ASIC are shown in Fig. 3.7. The HV supply powers the output driver stage. The analog circuits like the bandgap circuitry and bias current generation [6] are supplied by the MV rail allowing the design to use MV-rated components rather than using the HV rail and designing with HV-rated components. Using MV-rated components in design is area-efficient and achieves better matching [7].

The control signal to enable the bias current generation and driver stages is supplied by the LV stage. However we need to level shift the control signals to appropriate levels in order to have proper functionality of enabling and disabling the circuits as desired. Given the fact that all these three levels will not be available at the same time during the power-up phase, the chances of the tri-stated conditions and undefined nodes will be high and can lead to inadvertent activation or deactivation of the circuits. Hence there is a need for a robust biasing scheme that will overcome this challenge in the MSV design.

3.5 Typical Problems Observed in MSV Designs

The MSV circuits typically have at least a three-phase power-up as shown in Fig. 3.8. In phase I, HV is the only available rail. MV and LV are not available. In phase II, HV and MV rails are available but not the LV rail. Only in phase III, all the three rails are available. During phase I and II, the driver ASIC should ensure that there is no unintended activation or deactivation of the FETs especially the HV driver stages. The issue is addressed from simple level shifter issues.

Table 3.1 shows the different modules that are enabled at different phases of operation of the squib driver IC.

Level Shifters in MSV Designs: Turn *Off* Conditions
A cross-coupled level shifter, shown in Fig. 3.9, shifts the signal A_LV (~3.3 V) to A_MV (~5 V). This is valid only if LV level is available. When LV rail is not available during phase II where only the HV and MV rails are available, the output of this level shifter is undefined since MN1 and MN2 are cut off. This will make the level shifter completely tri-stated. Based on leakage of the transistors MP1, MP2, MN1 and MN2, both the output could also be either at MV rail or at ground.

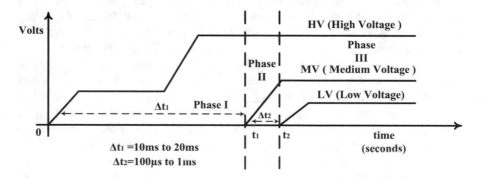

Fig. 3.8 Outputs from power management ASIC

Table 3.1 Power-up phases

Parameter	Phase I	Phase II	Phase III
Quiescent current (mA)	HV rail available. LV, MV not available Start-up reference voltage and start-up current generator are active Main bandgap, bias generator, driver circuitry *off*	HV, MV rails available. LV not available Start-up completed + bandgap and reference currents available, digital in reset, nPOR not released	HV, MV, LV rails available. Start-up completed + bandgap and reference currents + digital core out of reset, nPOR released + powerFETs activated when commanded by microcontroller

Fig. 3.9 LV to MV level
shifter tri-state condition

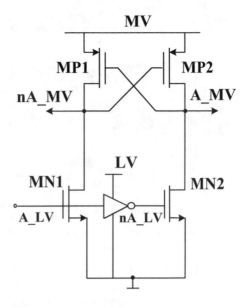

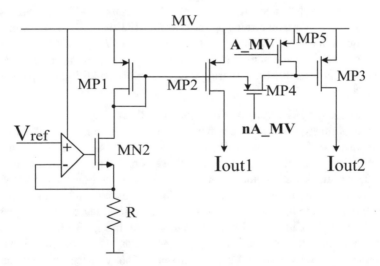

Fig. 3.10 Bias current generators

Let us assume a bias current generator shown in Fig. 3.10. The current through MP1 is V_{ref}/R. This current is mirrored as I_{out1} by MP2. The intended behaviour is to mirror an additional current of I_{out2} only by activating MP3. When the signal A_MV driving the gate of MP5 is at MV rail and the signal nA_MV driving the gate of MP4 is at 0 V, the gate of MP1 is connected to MP3.

In normal scenario like phase III, since the LV and MV rails are available, the outputs of the level shifter will be as desired.

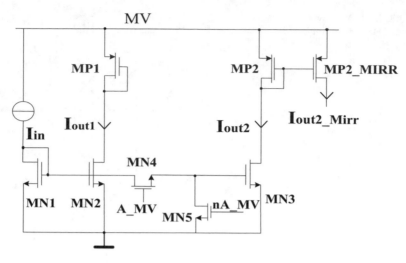

Fig. 3.11 Level shifter inadvertent turn on conditions

However due to the power sequencing, if the level shifter output is affected such that both signals A_MV and nA_MV are at 0 V, the gate of MP3 and MP1 will be pulled to MV level thereby turning *off* the current MP2 and MP3. The current V_{ref}/R will flow through MP4, MP5, MN2 and R. The currents I_{out1} and I_{out2} will be zero. In another scenario if the signal nA_MV is 0 and the signal A_MV is at MV, an unintended activation of MP3 will trigger I_{out2} current in addition to I_{out1} when it is not needed.

Level Shifters in MSV Designs: Turn *On* Situations
If the application requirements are to have the powerFETs turned *off*, the design has to ensure that the $V_{gs} = 0$ V both at power-up and steady state. Several designs can inadvertently activate the FETs on one of these conditions. Most of the issues are caused during the power-up phase. This is very critical, and designs in MSV needs novel circuit topologies in several applications to avoid inadvertent turn *on* of powerFETs to create low ohmic paths in the automotive that can endanger safety. Let us consider the situation with the current source. In Fig. 3.11, due to the level shifter behaviour in the absence of LV, if A_MV is at MV level and nA_MV is at 0, an unintended activation of MN3 can happen. This will lead to unintended or inadvertent current I_{out2} flowing through MN3 and MP2. If MP2 mirrors additional current sources, then there is too high current flow. Such a behaviour is not acceptable, and robust solutions need to be implemented.

High-Voltage Level Shifters
A similar scenario is applicable for the HV level shifter. The driver stages can have unintended activation or deactivation. The LV to HV level shifter converts a 3.3 V signal in LV domain to the 40 V HV domain. The BiCMOS process supports HV CMOS transistors whose V_{ds} rating is 40 V and V_{gs} rating is only 12 V. Hence the HV level shifter needs Zener clamps D_1 and D_2. The architecture is shown in Fig. 3.12a.

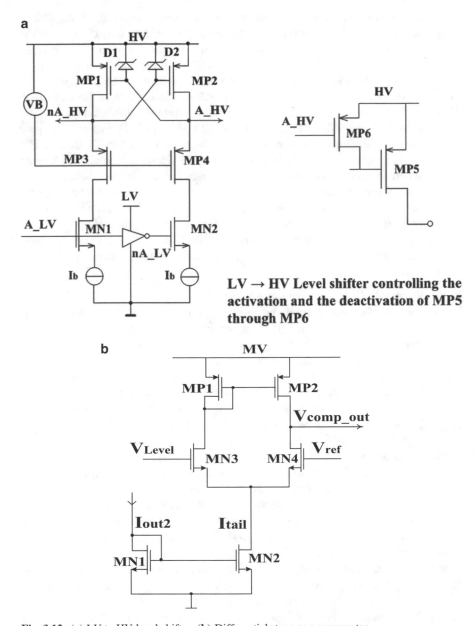

LV → HV Level shifter controlling the
activation and the deactivation of MP5
through MP6

Fig. 3.12 (**a**) LV to HV level shifter. (**b**) Differential stage as a comparator

This LV to HV level shifter is mainly needed to activate or deactivate the PMOS transistors connected to the HV rail. In this example, MP5 is turned *off* when the gate of MP6 is at HV-12 V. MP5 is turned *on* when the gate of MP6 is at HV level. Under normal conditions when both the LV and HV rails are present and the A_LV signal is 0, MN1 is *off*, and MN2 is *on*. Current I_b is mirrored from I_{out1} generated from the MV rail. This current flows from the HV rail through D_1, MP4 into MN2. This clamps A_HV node at HV-12 V, and the positive feedback makes nA_HV pulled up to the HV level. A_HV signal at HV-12 V turns *on* MP6 that pulls the gate of MP5 to HV rail and keeps it in *off* state. Similarly if A_LV = 1, then A_HV signal is at HV rail thereby turning *off* MP6. Gate of MP5 will be now set by its gate driver. When MV rail is not present, the bias current I_b is not present. When LV rail is not present, the inverter output is undefined thereby tri-stating the level shifter and impacting the functionality. In order to detect the inadvertent behaviour, the quiescent current of the driver ASIC from the HV rail in any phase of operation is a good measure of the indication of the tri-stated behaviour. Too low current indicates unintended deactivation, and too high current indicates unintended activation of circuits.

False Diagnostic Comparator Outputs
Due to the lack of the bias current, the comparators can react in the unintended way. These comparators could be a part of the monitoring circuits which when failed could cause the part to lock or remain in reset. An example is discussed below. The comparator shown below receives an input bias current from PMOS current source, I_{out2}, as shown in Fig. 3.12b. With the bias current in the order of pA, the comparator output is totally undefined. Theoretically it could be at MV/2 since it is an ideal impedance division between two resistors with GΩ of impedance. The V_{comp_out} node can be set at MV or 0, and the ASIC may be stuck at reset, or any driver stage can be inadvertently turned *on*.

To overcome the issue with the cross-coupled level shifters, a robust level shifting scheme has been proposed [8]. This is not area-efficient when a design has more than 50 level shifters, and the design complexity is high when the level shifter has to be designed to shift signals from LV to HV domain. Even if there is a scheme to ensure LV rail availability for the level shifter for phase II and III, the LV rail will not be available prior to the HV rail due to the requirements of the power management ASIC. To address this, the level shifters need to be functional independent of the LV rail. The bias current I_b has to be present from phase I to III to ensure correct functionality of the level shifter. State-of-the-art solutions [9] employ switching of the bias currents by using the Bandgap_Ready signal. The second bias block receives its input bias current based on the Bandgap_Ready signal. If this signal is 0, the Bias 1 serves as the input bias current to the second bias block, else the Bias 2 is used. Another biasing scheme uses a comparator-based decision to switch the bias currents [10, 11]. The current comparator decides if Bias 1 or Bias 2 will be provided as I_{bias_out}. Existing solutions use control signal-based decision regarding the biasing scheme. This suffers the disadvantage of expensive filtering to eliminate noise or suffers performance due to glitch in the control signal.

3.6   Elementary Cells for Automotive Circuits

In this section we look into some of the common circuit topologies used for automotive IC designs. Voltage and current limitation are one of the important design requirements in the automotive ICs. This is needed to avoid potential destruction to the IC arising out of power dissipation or impact ionization or avalanche breakdowns [12]. Traditional cells are reported in several analog literatures [13, 14] where the circuits can be built in order to limit the voltage or current.

In Fig. 3.13a, the NMOS transistor can be biased with a resistor and a Zener diode such that a voltage limitation is obtained. HV level can even be 40 V. However if the gate voltage V_{bias} is clamped, for instance, to 6 V, then V_{out} is limited to ~ 5 V. It is important to ensure that the transistor rating is 40 V, i.e. the M1 transistor can withstand a 40 V delta voltage between its drain and source terminals and V_{gs} of 12 V. Similarly in Fig. 3.13b, the current mirror MP1 – MP2 delivers the I_{bias} current to the load when S_1 is activated. When the output V_{out} is shorted to ground, the current that flows out of the switch S_1 is I_{bias}. This current mirror now functions as a current limiter as the current cannot exceed I_{bias} even under this fault condition. It is important to ensure that, when HV is at 40 V and the V_{out} is at 0 V, the V_{sd} of the PMOS FET MP2 should be able to handle to 40 V condition, and if I_{bias} is in the mA range, the junction temperature of the FET MP2 device [15] will increase and the device should be able to withstand the thermal effect without destruction.

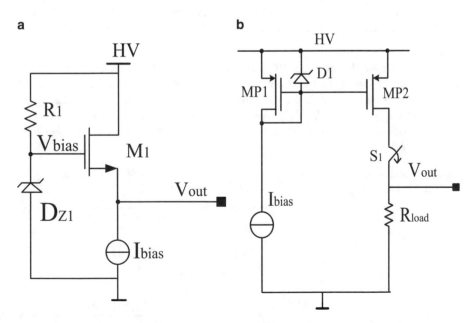

Fig. 3.13 (**a**) Source follower as a voltage limiter. (**b**) Current mirror as a current limiter

3.7 Maximum Voltage and Current Selector Circuits

A robust biasing scheme can be developed by using the voltage and current selector circuits. When two NMOS transistors MN1 and MN2 are connected in the common source configuration, the maximum voltage is selected as the dominant supply. This is shown in Fig. 3.14. If MN1 and MN2 threshold voltage is 1 V with the same W/L ratio, the source of the FETs will be at 4 V as it is defined by the transistor with the high overdrive. In case the transistors have similar overdrive, then the dominant path can be altered by changing the W/L of the transistors. The one with a higher W/L ratio is the stronger one and becomes the dominant path for the current flow. Similarly MN1 and MN2 can be replaced by two diodes. The diode with the anode at the maximum level will be the dominant supply.

A maximum current selector (MCS) selects the maximum of the two currents [16, 17]. This is shown in Fig. 3.15. The current selector receives two currents I_R and I_S. The maximum current selector (MCS) monitors the difference current $|I_S - I_R|$ through MP4. MP7 mirrors the current I_R. MN5 becomes the current adder thereby providing the maximum current between I_S and I_R.

Phase I Analysis

The circuit can be analysed by considering the phase I operation where I_S is present and I_R is not present. In phase I operating mode, $I_S = 2$ μA and $I_R = 0$ μA. MN2 current is 2 μA. Since $I_R = 0$ μA, the current through MP1 is 0. Hence the current through MP4 is I_S. Since $I_R = 0$ μA, the current through MP6 is 0. As a result, the current through MN5 is $I_S = 2$ μA. Thus the maximum current between I_S and I_R is delivered. These selector circuits will be extensively used to develop the biasing scheme for the drivers operating in a multiple supply voltage (MSV) system.

Phase II–III Analysis

During the phase II and phase III mode of operation, the reference current I_R is available in addition to I_S. It is normally higher than I_S since several modules in the driver ASIC operate with a 10 μA reference current. Under this condition, the current through MP1 is 10 μA, whereas MN2 cannot sink more than 2 μA. This situation will force transistor MP4 to source only 2 μA by making the transistor

Fig. 3.14 Maximum voltage selector

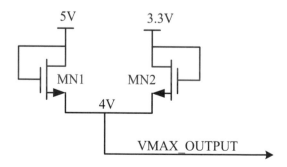

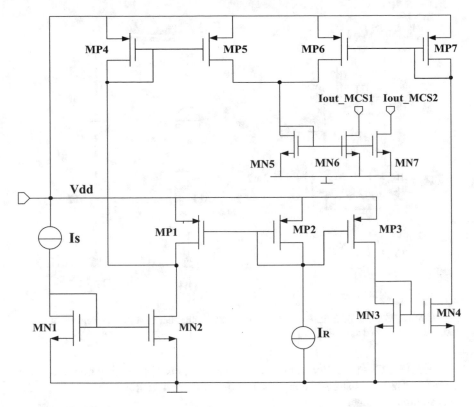

Fig. 3.15 Maximum current selector [17]

operate in the linear region. As a result, the current through transistors MP4 and MP5 is zero. MP7 and MP6 mirror 10 μA. As a result, the current through MN5 is 10 μA which is the maximum of the two currents 2 and 10 μA.

3.8   Differential Voltage and Current Amplifiers

The backbone of current sensing structures and diagnostic circuitry is the differential amplifiers. Based on requirements like gain, bandwidth, input, output impedances, etc., two types of differential structures would be required. One is based on voltage input and the other is on the current input. This is shown in Fig. 3.16a, b, respectively. For high-gain, high-input, high-output impedances at the cost of low bandwidth, voltage input differential stage is used. Figure 3.16a is the traditional differential to single-ended, voltage input-based topology explained in almost all electronics text books. When low-gain, low-input, low-output impedances and high bandwidth are required, current input differential stage is preferred. This is shown

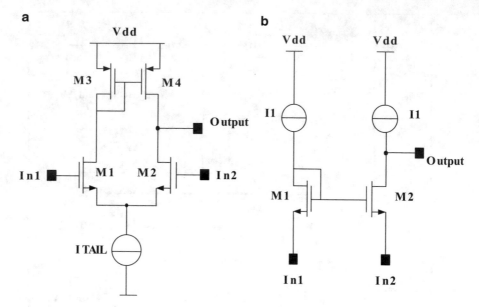

Fig. 3.16 (**a**) Voltage input differential to single-ended amplifier [13]. (**b**) Current input differential to single-ended amplifier [13]

in Fig. 3.16b. Change in the signal I_{n1} appears as a change in the gate voltage for transistor M2. Transistor M2 converts the equivalent differential voltage of $I_{n1} - I_{n2}$ to the output voltage.

3.9   Summary and Conclusion

In this chapter, we discussed about the two current sensing techniques. In multiple supply voltage circuits, the level shifter behaviour is very critical as it can create unintended activation or deactivation of circuits. The biasing schemes are critical and have to be robust to ensure functionality independent of the different phases of power sequencing. The bias current should be present at every phase of operation. The voltage and current selector circuits can be exploited to achieve robust biasing schemes and preventing unintended or inadvertent activation and deactivation of circuits.

References

1. T. Regan, Current sense circuit collection, *Linear Technology*, Application Note 105, Dec 2005
2. *Current Loop Application Note*, CLAN1495, B&B Electronics, 1995

3. *User Configurable Airbag IC*. L9678, L9687-S, Datasheet, DocID025869, Rev 3.0, STMicroelectronics, May 2014
4. *2-Ch Squib Driver For Airbags*. TPD2004F, Datasheet, Toshiba, Oct 2006
5. *4 Channel Squib Driver*. E981.17, Datasheet, QM-No. 25DS0073E.00, Elmos, Aug 2011
6. B. Razavi, *Design of Analog CMOS Integrated Circuits* (McGraw-Hill, New York, 2001)
7. M.J.M. Pelgrom, A.A.D.C.J. Duinmaijer, A.P.G. Welbers, Matching properties of MOS transistors. IEEE J. Solid State Circuits **24**(5), 1433–1440 (1989)
8. M.H. Kim, Level shifter having single voltage source, U.S. Patent US7, 683,667B2, 23 Mar 2010
9. C.Y. Chen, System and method for breakdown protection in start-up sequence with multiple power domains, U.S. Patent US 7,391,595B2, 24 June 2008
10. B. Sharma et al., Avoiding excessive cross-terminal voltages of low voltage transistors due to undesirable supply-sequencing in environments with higher supply voltages, U.S. Patent US 7,176,749B2
11. Y. Date et al., Current driver, U.S. Patent US 7,466,166B2, Dec 2008
12. C. McAndrew, Y. Tsividis, *Operation and Modelling of the MOS Transistor* (Oxford University Press, New York, 2012)
13. V. Ivanov, I. Filanovsky, *Operational Amplifier, Speed and Accuracy Improvement* (Kluwer, Boston, 2004)
14. E.A.M. Klumperink, *Transconductance based CMOS Circuits: Generation, Classification and Analysis*, PhD Thesis, University of Twente, Enschede, The Netherlands, 1997
15. P.L. Hower, Safe operating area – a new frontier in LDMOS design, in *Proceedings 14th International Symposium on Power Semiconductor Devices and ICs*, Santa FE, New Mexico, June 2002, pp. 1–8, doi: https://doi.org/10.1109/ISPSD.2002.1016159
16. E. Seevinck, R.J. Wiegerink, Generalized translinear circuit principle. IEEE J. Solid State Circuits **SC-26**(8), 1098–1102 (1991)
17. J. Madrenas et al., Self-controlled 4-transistor low power min-max current selector. Int. J. Electron. Commun. (AEü) **63**, 871–876 (2009)

Chapter 4
High-Side Current Regulation and Energy Limitation

The current regulation loop drives squibs that are used to inflate multiple airbags in cars. Squibs are electrically *R-L-C* networks with the resistance ranging from 1 Ω to 8 Ω. The inductance range varies from 1 μH to 3 mH and the capacitive loads range from 22 nF to 220 nF. The high-voltage LDMOS transistors are used as the HS and LS FETs. The specified currents are minimum 1.2 A for 2 ms, minimim 1.75 A for 0.5 ms or 0.7 ms and an optional mode of minimum 1.2 A for 1.5 ms. In this chapter, the advantages and disadvantages of current sense circuit with sense resistor and senseFET approaches are compared. Based on this comparison, this chapter shows how the targeted current regulation is achieved for the HS driver by using the sense resistor-based approach along with the energy limitation circuit for inadvertent deployment protection.

4.1 Sense Resistor and SenseFET Comparison

In sense resistor current sensing scheme, the aluminium metal sense resistor is used since we know that a poly resistor is not effective for high current. This resistor is placed on the drain of the LDMOS powerFET. In the senseFET current sensing scheme, the senseFET is connected in parallel to the powerFET. This is shown in Fig. 4.1a, b. Table 4.1 shows the advantages and disadvantages.

Table 4.1 shows the advantages between the sense resistor- and senseFET-based current sensing scheme. Since the accuracy is very important for the current regulation of the HS_FET, metal resistor-based current sensing is the best choice. However it suffers the disadvantage of self-heating and needs some compensation mechanism to achieve the targeted current. The $R_{\text{ds\_on}}$ from Vdd to the load due to the metal resistance is also a disadvantage but can be overcome with appropriate measures. Vdd is the high voltage (HV) rail.

© Springer International Publishing AG 2018
S.N. Easwaran, *Current Sensing Techniques and Biasing Methods for Smart Power Drivers*, https://doi.org/10.1007/978-3-319-71982-5_4

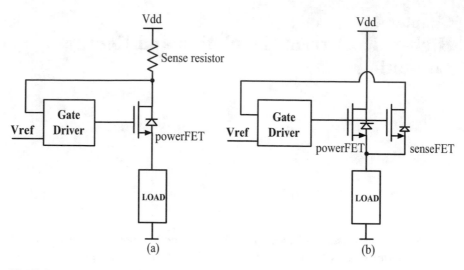

Fig. 4.1 Current sensing schemes. (**a**) Sense resistor scheme. (**b**) SenseFET scheme

Table 4.1 Sense resistor, senseFET comparison

Spec parameter	Target	Notes
Placement	Metal resistor placed in series with the power FET	Parallel with the powerFET
Current regulation accuracy	More accurate	Less accurate
Feedback – output sampling scheme	Shunt feedback	-NA-
R_{ds_on} impact from Vdd to load	High since metal resistor resistance adds to the FET resistance	No impact since FET resistance is the only factor
Mismatch with regard to the reference	Medium. Needs better layout techniques	Medium. Needs better layout techniques
Thermal behaviour	High and rapid self-heating on the sense element	Negligible self-heating on the sense element
Current regulation/ application levels	Current <3 A	Current >3 A

4.2   High-Side (HS_FET) Current Regulation

An aluminium metal sense resistor approach on the drain of the HS_FET is used to regulate the current. The VZx supply varies from 18 to 33 V. The metal resistor r_1 is used. The voltage drop across r_1 is compared with a reference voltage across metal resistor r_2, generated through I_{ref} to regulate the current. This is shown in Fig. 4.2. The design steps towards the accomplishment of the current regulation of the HS_FET are described below.

Fig. 4.2 HS_FET current
regulation with a metal
sense resistor

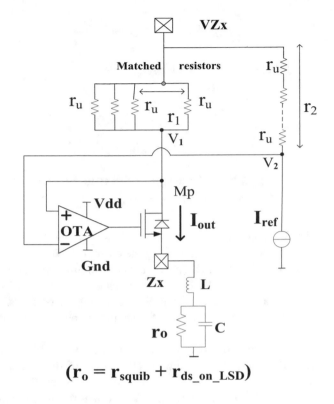

Current Regulation Principle and Critical Design Values

The circuit shown in Fig. 4.2 regulates the current to a value proportional to the reference current and the ratio of the two resistors r_2 and r_1. To understand the basic regulation principle, let us assume that the offset error due to the OTA is negligible. The OTA forces its positive input V_1 to the reference voltage V_2 which is set by I_{ref} and r_2. The value of r_1 is chosen such that there is sufficient voltage drop across the resistor when the minimum deployment current of 1.2 A flows through the resistor even though the targetted current would be approximately 1.4 A. This is needed to ensure that the sensed voltage is dominant and is not masked by the offset of the OTA. If the sensed voltage has enough margin for the 1.2 A current level, it obviously guarantees more than sufficient margin for the 1.75 A current level (targetted current level is 2 A). The values for r_2 and I_{ref} are chosen based on the amount of headroom needed for the OTA from the VZx rail. Based on these factors, r_1 is set to 120 mΩ, and r_2 is set to 7.2 Ω. I_{ref} is set to 23.3 mA for 1.4 A current and 33.3 mA for 2 A current regulation levels.

$$I_{ref} = \frac{VZx - V_2}{r_2} \qquad (4.1)$$

$$I_{\text{out}} = \frac{\text{VZx} - V_1}{r_1} \tag{4.2}$$

Since V_1 equals V_2, we get

$$I_{\text{out}} = \frac{r_2}{r_1} I_{\text{ref}} \tag{4.3}$$

4.3 $R_{\text{ds\_on}}$ Specification, Thermal Simulations and PowerFET Dimension

The $R_{\text{ds\_on}}$ between VZx and Zx pins is specified to be 1 Ω maximum. The metal resistor value chosen for sensing the current is 120 mΩ. The $R_{\text{ds\_on}}$ of the HS_FET is chosen to be 500 mΩ. This ensures the nominal value is 620 mΩ with a maximum resistance of 800 mΩ. In order to reduce the routing resistance, copper traces are used. It has low sheet resistance and hence is the preferred trace for routing. Initial floor plan is performed to estimate the layout area of the FETs along with its metallization.

The maximum supply rail VZx can be 33 V, and Zx can be at 0 V due to the short-to-ground fault condition. The choice of the output current is controlled by software in the airbag control unit (ACU). When the HS_FET is activated, the current of 2 A flowing through the FET will result in a power dissipation of 66 W for 0.5 or 0.7 ms. Similarly for the 1.4 A current setting, the power dissipation will be 46 W for 2 ms. This power dissipation will result in a rapid increase in the junction temperature T_j of the HS_FET. Thermal simulators like FloTHERM [1] are used to analyse the temperature rise and the temperature profile. Through the thermal simulations, we ensure that the area of the HS_FET is not undersized but sufficient enough to handle the heat dissipation. It does ensure that the FET is not oversized for thermal dissipation. In other words, it ensures that the HS_FET is within the thermal SOA (safe operating area) [2–4] limits. The thermal SOA is measured by using the critical temperature T_{crit}. This is defined as the temperature above which the HS_FET gets completely conductive and cannot be turned *off* even if the $V_{\text{gs}} = 0$ V. T_{crit} is determined by the process technology. For TI BiCMOS process, this is 550 °C for short activation periods less than 3 ms.

In addition to the thermal SOA analysis of the HS_FET, the additional objective of the thermal simulations is to also understand the temperature rise on the locations adjacent to the HS_FET. This will help to ensure that the thermally sensitive elements of the current regulation loop like the OTA are not placed adjacent to the HS_FET.

Thermal Simulation for T_j Rise with Multiple Drivers in a Driver IC

The junction temperature (T_j) increase is estimated from the ambient temperature (T_a) and the junction to ambient thermal resistance denoted as Theta J_A (θ_{JA}). The junction to package thermal resistance is negligible for deployment durations less than 3 ms. The JEDEC (Joint Electron Device Engineering Council) board conditions are assumed. θ_{JA} is assumed to be 23.6 C/W for the assumed package and die size. The power dissipation across the HS_FET is $V_{ds}.I_d$ where V_{ds} is the drain to source voltage drop across the HS_FET and I_d is the drain current flowing through the HS_FET. T_j is then calculated by the formula shown below.

$$T_j = T_a + \theta_{JA}V_{ds}I_d \qquad (4.4)$$

Thermal simulation is performed on the four-channel driver with all the HS_FETs, handling energy of 48 mJ or 98 mJ simultaneously. Thermal simulations are performed for both energy levels. This way the thermal cross talk between the HS_FETs and the thermal cross talk from each HS_FET to other modules like bandgap, bias current generator, diagnostic circuitry and digital controller in the driver IC is checked. The placement of the FETs and the locations to be monitored are shown in Fig. 4.3a, b. The HS_FETs and LS_FETs are the dominant portion of the chip. These FETs are placed adjacent to their bond pads as the routing resistance

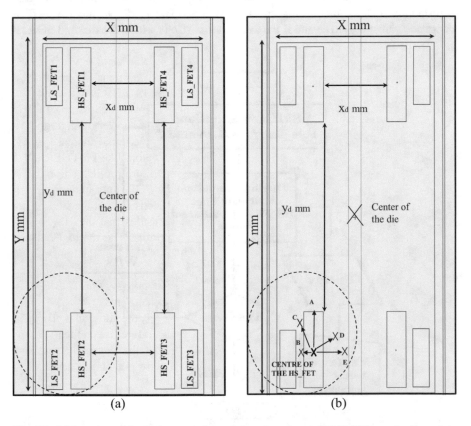

(a)   (b)

Fig. 4.3 (a) Location of the deployment channels based on pinout. (b) HS_FET monitoring points T_j rise

should be reduced due to the high current flow. The chip is X mm wide and Y mm long. The vertical separation between two channels in the vertical direction is Y_d mm, and the separation is X_d mm in the horizontal direction.

The ambient temperature is set to 115 °C. However an airbag control unit runs several diagnostic tests on the driver IC prior to a deployment event. These diagnostic tests will dissipate additional power and will increase the on-chip junction temperature above ambient T_a. FloTHERM results indicate that the on-chip junction temperature rises to 135 °C. Hence T_a for the device prior deployment start was assumed to be 135 °C. The expected output from FloTHERM is to provide the junction temperature rise within the HS_FET and its neighbouring locations during deployment. As shown in Fig. 4.3b, the monitoring points of interest from the centre of the HS_FET are identified.

Flow Chart
The flow chart in Fig. 4.4 shows how the SPICE simulator and thermal simulator are effectively utilized to arrive at the optimal design. The aspect ratio (W/L) of the HS_FET is first estimated with SPICE simulator for R_{ds_on} (~500 mΩ). This (W/L) is provided to the thermal simulator along with inputs related to 48 or 98 mJ energy information and the monitoring areas of interest. The T_j rise in the HS_FET is checked against T_{crit} to ensure if the temperature rise is within the thermal SOA. If

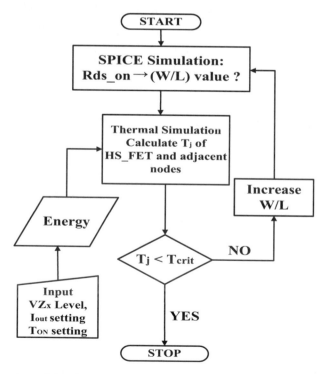

Fig. 4.4 Flow chart of electrical and thermal flow

$T_j > T_{crit}$, thermal SOA is violated, and W/L of the HS_FET will be increased to ensure that thermal SOA is met. The temperature rise in the neighbouring locations of the HS_FET is also obtained from the thermal simulations in order to ensure that sensitive circuits like the OTA are not affected at peak temperatures of 400 °C. After estimating the W/L of the HS_FET with SPICE simulator, this value is provided to the thermal simulator along with the monitoring points of interest from the centre of the HS_FET. The HS_FET T_j rise is checked if it is within the thermal SOA.

The temperature rise in the neighbouring locations of the HS_FET is also obtained from the thermal simulations in order to ensure less thermal impact to thermally sensitive components in the OTA. If the $T_j > T_{crit}$, then the thermal SOA is violated, and the W/L of the HS_FET will be increased to ensure that the thermal SOA is met. The R_{ds_on} will reduce, but it will not have much impact to the current regulation performance as the HS_FET will be in the saturation region of operation when in regulation.

Thermal Simulation Results

The thermal simulation results are represented in two formats. In one format, the transient thermal behaviour of the HS_FET and its neighbouring locations from ambient temperature is reported. This is shown in Fig. 4.5a, b. In the second format, thermal analysis for the whole driver IC is represented by isothermal plots. This is shown in Fig. 4.6a, b for both the deployment current settings and deployment times.

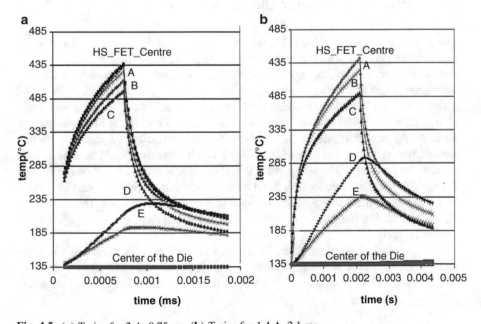

Fig. 4.5 (a) T_j rise for 2 A, 0.75 ms. (b) T_j rise for 1.4 A, 2.1 ms

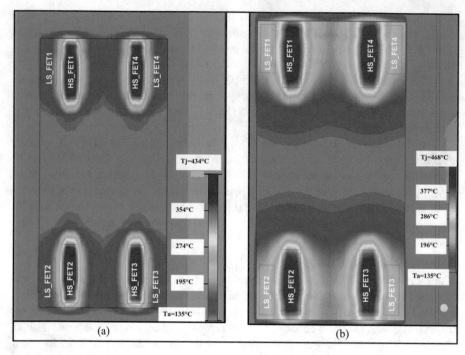

Fig. 4.6 (**a**) Isothermal plot, T_j, rise for 2 A, 0.75 ms. (**b**) Isothermal plot, T_j, rise for 1.4 A, 2.1 ms

4.4 Metal Sense Resistor Dimension

The metal sense resistor value [5] is chosen to be 120 mΩ. However the width of the metal resistor is critical and should be wide enough to handle the 2 A current flow. It should neither be too low nor too high since narrow trace results in electromigration, whereas the very wide trace is not area- and cost-efficient. If the metal resistor is dimensioned to handle the maximum current of 2.14 A, the resistor will also handle the 1.4 A current level without issues. In addition to the regulated currents, the peak currents can be higher when fault conditions like short to ground are encountered on the source of the HS_FET. The resistor should also be able to handle these currents. The width of the metal resistor is chosen based on the current density. The current density is defined as the minimum current that can be carried by a 1 µm metal trace above which the trace will melt and fuse open. There are three classifications for the current density based on the current duration [6]. This is shown in Table 4.2.

If J_{DC} is used to calculate the metal width, then for 2 A level, the metal width has to be 2000 µm. This is extremely large area and is not cost-effective even for a single driver. To optimize the metal width, the repetition rate of the deployment

Table 4.2 Current density for metal traces

Types of current density	Notation	Thumb rule mA/μm	Application
Peak	J_{PEAK}	50	ESD currents
RMS	J_{RMS}	10	PWM currents
DC	J_{DC}	1	DC currents

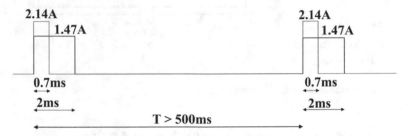

Fig. 4.7 Deployment pulse repetition rate

current pulse should be considered. In Fig. 4.7, the repetition rate is higher than 500 ms with the duty cycle being 700 μs for the 2.14 A maximum current level or 2 ms for the 1.47 A maximum current level. J_{RMS} is apt for the pulse width modulation (PWM)-based current flow like in switching converters where the switching period is in the order of 1–2 μs. The situation considered here indicates the current flow for a short duration through the metal resistor and a long duration where there is no current flow. J_{RMS} is still a nonoptimal value for area-efficient width of the metal trace. J_{PEAK} is intended for short pulses in "ns" range which is for ESD currents. J_{PEAK}- and J_{RMS}-based calculations will result in 40 and 200 μm trace widths. The width of 40 μm was chosen as this was area-efficient.

Metal Resistor and Copper Layout
A unit aluminium metal resistor r_u is used to form several parallel and series segments of the unit resistor r_u. Each unit resistor is 10 μm wide. Here the 120 mΩ resistor is formed by a parallel combination of 600 mΩ unit resistors. The parallel structure provides an effective width of 50 μm. This way the minimum width requirement of 40 μm based on the J_{PEAK} calculations are fulfilled. Similarly 12, 600 mΩ unit resistors connected in series form the 7.2 Ω resistor to generate the reference voltage for the OTA. The width of the 7.2 Ω resistor is 10 μm which ensures the reference current I_{ref} of 33.3 mA can flow through without electro-migration issues. Each unit metal resistor r_u is connected with copper. Multiple vias are used to connect the aluminium metal resistor to the copper trace. It is ensured that there are more than sufficient vias to handle this high current. The copper interconnect does ensure very less routing resistance and ensures negligible error to the resistor ratio of 60. This is typically verified with the parasitic R-C extracted simulations. This type of layout structure shown in Fig. 4.8 achieves good matching [7] so that any process gradient will have less impact to the ratio.

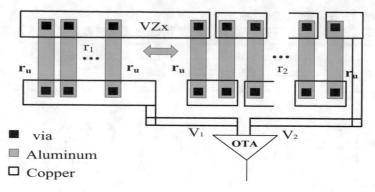

Fig. 4.8 Aluminium metal resistor layout scheme

4.5 Current Regulation and Small-Signal Analysis

The regulated current from the HS_FET including the offset of the OTA is shown in the equation below.

$$I_{\text{out}} = I_{\text{ref}} \left(\frac{r_2(1 + \Delta t)}{r_1(1 + \Delta t)} \right) + \frac{\Delta V_{\text{offset}}}{r_1(1 + \Delta t)} \tag{4.5}$$

The metal resistors have positive temperature coefficient, and this will cancel out the temperature variation. Since the current regulation relies on feedback principle with the load on the Zx pin being an *R-L-C* network, the small-signal model is necessary to understand the stability of the loop. The OTA is assumed to be a two-stage amplifier with one high impedance node. The OTA is referenced with reference to ground. The powerFET is referenced with reference to Zx, and hence it is necessary to use a source follower that is referenced to Zx pin. It is important to note that the gain from the powerFET stage of the current regulation loop is set by r_o and r_1. For our analysis, we refer r_1 here as R_{sense}. Intuitively the powerFET along with r_o is a degenerated stage whose transconductance is determined by $1/r_o$ since $r_o > 1$. The gain is now set by r_o and R_{sense}. Since R_{sense} is very low, this stage becomes an attenuation stage. Hence a very high gain has to be generated at the OTA such that the overall gain is around 90 dB in order to achieve $\pm 10\%$ accuracy for the regulated current. The second pole is determined totally by the *R-L* load at the output. Source follower M_f is added to drive the powerFET M_p. Fig. 4.9 shows the AC open-loop simulation method. Figure 4.10 shows the small-signal model.

This becomes a two-pole system with poles lying closer to each other. The OTA is represented as a two-stage amplifier with first-stage transconductance g_{m1} and output impedance R_a. The second-stage transconductance is g_{m2} with output

Fig. 4.9 Open-loop
simulation set-up

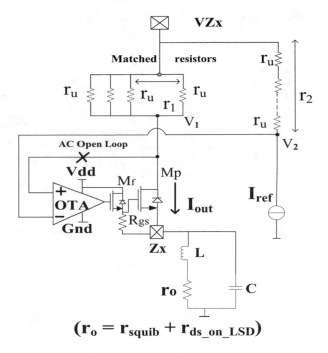

$$(r_o = r_{squib} + r_{ds\_on\_LSD})$$

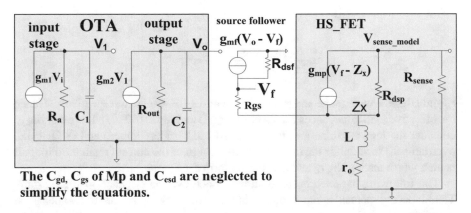

The C_{gd}, C_{gs} of Mp and C_{esd} are neglected to
simplify the equations.

Fig. 4.10 Small-signal model of the current regulator without frequency compensation

impedance R_{out}. The HS_FET transconductance is g_{mp}. The OTA sets the dominant
pole ($R_aC_1 \ll R_{out}C_c$); the second pole is set by r_o and L. Table 4.3 shows design
choices indicating design components that can be changed to stabilize the loop.

Table 4.3 Design parameters and choices

Amplifier stages	Parameter	Degree of freedom	Condition
1st (input) Stage_OTA	$g_{m1}R_a$	√	Open-loop gain, stability
2nd (output) Stage_OTA	$g_{m2}R_{out}$	√	Open-loop gain, stability
Source follower	g_{mf}	√	Open-loop gain, stability
PowerFET	g_{mp}	×	R_{ds_on}, thermal
Sense resistor	R_{sense}	×	R_{ds_on}
Load	r_o, L	×	Squib resistance, wiring inductance defined

For L in μH, $r_0 > 1\ \Omega$, the effective transconductance of the HS_FET is r_0.

$$\left(\frac{g_{mp}}{1+g_{mp}r_o}\right) \sim r_o \text{ if } g_{mp}r_o \geq 1$$

$$H(s) \sim \frac{g_{m1}R_a g_{m2}R_{out}R_{sense}}{r_o\left[(1+sR_{out}C_2)\left(1+s\frac{L}{r_o}\right)\right]}$$

$$f_{p1} = \frac{1}{2\pi R_{out}C_2}; \quad f_{p2} = \frac{r_o}{2\pi L};$$

$$f_\tau \simeq \frac{1}{2\pi}\sqrt{\frac{g_{m1}R_a g_{m2}R_{sense}}{LC_2}};$$

(4.6)

r_o and L have a wide range and can be 1–8 Ω and 1 μH–100 μH, respectively. Hence the effective load impedances can be (1 Ω, 1 μH) and (8 Ω, 100 μH). It is essential to consider the load combinations like (1 Ω, 100 μH: underdamped) and (8 Ω, 1 μH, overdamped) to stabilize the loop. After the design of the current regulation loop, it is also worth testing the r_o and L combination of 1 Ω and 1 μH.

The following parameters are assumed to achieve the targeted open-loop gain. The expected parasitic capacitances are in the 400 fF on an uncompensated network.

$$g_{m1} = 20\mu S, R_a = 20K\Omega, g_{m2} = 30\mu S, R_{out} = 60G\Omega, C_1 = C_2 = 400fF,$$

$$g_{mf} = 200\mu S,$$

$$R_{dsf} = 3M\Omega, R_{gs} = 100K\Omega, g_{mp} = 3S, R_{dsp} = 5\Omega, R_{sense} = 120m\Omega, r_0 = 2\Omega,$$

$$L = 100\mu H.$$

The poles (f_{p1}, f_{p2}), zero (f_{z1}) and the unity gain bandwidth (f_τ) are calculated as per Eq. (4.6).

$$H(s) = \frac{(20.10^{-6})(20.10^3)(30.10^{-6})(60.10^9)(0.12)}{2} = \frac{86400}{2} = 43200$$

$$\Rightarrow A_{dc} = 20\log_{10}(43200) = 92 \text{ dB}.$$

$$f_{p1} = \frac{1}{2\pi(60.10^9)(400.10^{-15})} = 6.63 \text{ Hz}$$

$$f_{p2} = \frac{2}{2\pi(100.10^{-6})} = 3.18 \text{ kHz}$$

$$f_\tau = \frac{1}{2\pi}\sqrt{\frac{(20.10^{-6})(20.10^3)(30.10^{-6})(60.10^9)(0.12)}{(100.10^{-6})(400.10^{-15})}} = 30 \text{ kHz}$$

Frequency Compensation

An ideal scenario is to have a single-pole system that is always stable. Theoretically if we can introduce a zero at the location of the second pole, pole-zero cancellation occurs thereby making the system completely stable. The small-signal model [8] without frequency compensation is shown in Fig. 4.11. The loop is stabilized by using the capacitor C_c and zeroing resistor R_{z1}. The poles f_{p1}, f_{p2} and zero f_{z1} and the unity gain bandwidth (UGB) are derived as shown below. To improve the stability for R-L-C loads where the L is on the higher side and R and C are on the minimum side, the capacitor C_d is added as to improve the damping in the overall transfer function. In this way a robust frequency compensation scheme has been developed and supports the wide R-L-C load range. The strategy is shown in Fig. 4.12. The open-loop AC parameters are derived below.

For L in μH, $L \ll R_{out}C_c r_0$, the transfer function is given in Eq. (4.7). A_{dc} is unchanged.

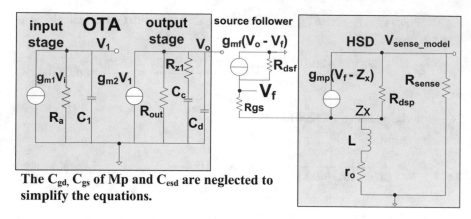

The C_{gd}, C_{gs} of Mp and C_{esd} are neglected to simplify the equations.

Fig. 4.11 Small-signal model of the current regulator with frequency compensation

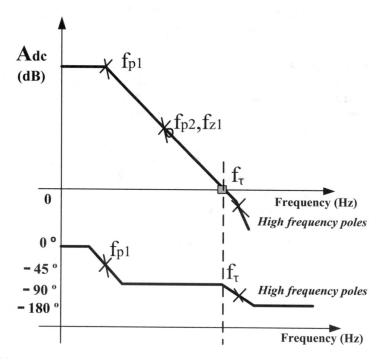

Fig. 4.12 Bode plot concept for pole-zero cancellation

$$H(s) = \frac{g_{m1}R_a g_{m2}R_{out}\left(\dfrac{g_{mp}}{1+g_{mp}r_o}\right)R_{sense}(1+sR_{z1}C_c)}{\left[1+sR_{out}\ C_c + s^2\left(LR_{out}C_c\left(\dfrac{g_{mp}}{1+g_{mp}r_o}\right)\right)\right]}$$

When $r_o > 1\Omega$, $\dfrac{g_{mp}}{1+g_{mp}r_o} \sim r_o$

$$H(s) = \frac{g_{m1}R_a g_{m2}R_{out}R_{sense}(1+sR_{z1}C_c)}{r_o\left[1+sR_{out}C_c + s^2\left(\dfrac{LR_{out}C_c}{r_o}\right)\right]}$$

$$f_{p1} = \frac{1}{2\pi R_{out}C_c};$$

$$f_{p2} = \frac{r_o}{2\pi L}\quad ;$$

$$f_{z1} = \frac{1}{2\pi R_{z1}C_c}\quad ;$$

$$f_\tau = \frac{g_{m1}R_a g_{m2}R_{sense}}{2\pi r_o C_c};\tag{4.7}$$

In order to achieve an open-loop gain of 90 dB and target a single pole at the output of the OTA, the following parameters are chosen.

$g_{m1} = 20\mu S, R_a = 20K\Omega, g_{m2} = 30\mu S, R_{out} = 60G\Omega, C_c = 4pF,$
$R_{z1} = 10M\Omega, g_{mf} = 200\mu S,$
$R_{dsf} = 3M\Omega, R_{gs} = 100K\Omega, g_{mp} = 3S, R_{dsp} = 5\Omega, R_{sense} = 120m\Omega, r_0 = 2\Omega,$
$L = 100\mu H.$

Open Loop Gain, Poles and Zeros, UGB after Compensation

$$A_{dc} = \frac{(20.10^{-6})(20.10^3)(30.10^{-6})(60.10^9)(0.12)}{2} = \frac{86400}{2} = 43200$$

$$=> A_{dc} = 20\log_{10}(43200) = 92\text{dB}$$

$$f_{p1} = \frac{1}{2\pi(60.10^9)(4.10^{-12})} = 663\ \text{mHz}$$

$$f_{p2} = \frac{2}{2\pi(100.10^{-6})} = 3.183\ \text{kHz}$$

$$f_{z1} = \frac{1}{2\pi(10.10^6)(4.10^{-12})} = 3.98\ \text{kHz}$$

$$f_\tau = \frac{(20.10^{-6})(20.10^3)(30.10^{-6})(0.12)}{2\pi(2)(4.10^{-12})} = 28.64\ \text{kHz}$$

PhaseMargin

$$\Phi_m = 180^\circ - \arctan\left(\frac{f_\tau}{f_{p1}}\right) - \arctan\left(\frac{f_\tau}{f_{p2}}\right) + \arctan\left(\frac{f_\tau}{f_{z1}}\right)$$

$$\Rightarrow \Phi_m = 180^\circ - \arctan\left(\frac{28.64\text{K}}{0.633}\right) - \arctan\left(\frac{28.64\text{K}}{3.183\text{K}}\right) + \arctan\left(\frac{28.64\text{K}}{3.98\text{K}}\right) = 90^\circ$$

The Bode plot for the uncompensated and compensated current regulation loop is shown in Fig. 4.13. It can be seen that the uncompensated loop has two poles within the UGB such that the phase margin is 0°. The compensated loop cancels the

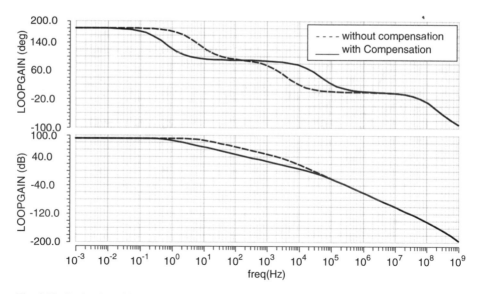

Fig. 4.13 Bode plot with and without frequency compensation

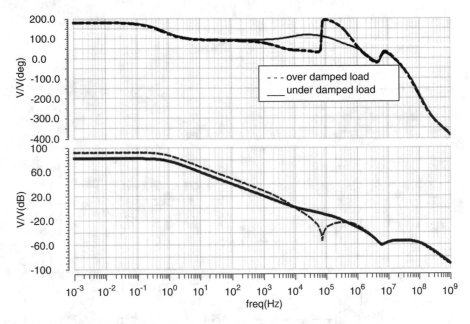

Fig. 4.14 Bode plot for stabilized loop for overdamped and underdamped loads

second pole f_{p2} with the zero f_{z1}. Figure 4.14 shows the Bode plot for the stabilized loop for the overdamped and underdamped loads. The schematic of current regulation is shown in Fig. 4.15. Figure 4.16 shows the open-loop Bode plot simulations with the implemented current regulation loop for both the overdamped and underdamped loads. Figure 4.17 shows the transient simulations for 1.4 and 2 A settings for both overdamped and underdamped loads. The deployment time was extended to 3 ms in order to check the free-wheeling behaviour for a 3 mH inductive load. Figure 4.18 shows the 1.4 A, 3 ms current regulation behaviour and the internal free-wheeling path provided by the HS_FET. This avoids the need for external Schottky diodes.

The resistor R_{gs} in the source follower ensures $V_{gs} < V_{th}$ to avoid unintended activation of M_p during unpowered conditions. The diode D_2 is a Zener protection for the gate-source junction of M_p. The parasitic diode D_{par} clamps the gate of M_p to -0.7 V when Zx flies negative due to the inductive kick [9] while M_p is turning off. This positive V_{gs} on M_p circulates the free-wheeling current and avoids external Schottky diodes. The test for the internal free-wheeling path was tested by using a special mode for test where the deployment activation time can be set to 3 ms, and a 3 mH coil was used to check the behaviour.

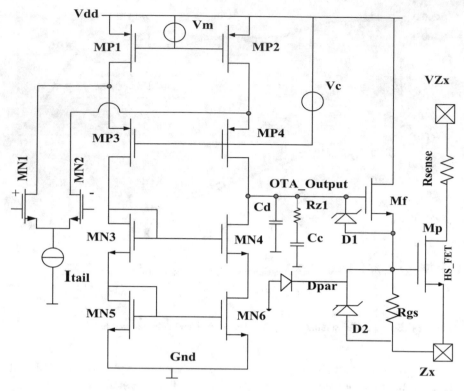

Fig. 4.15 OTA with HS_FET and the driver stage

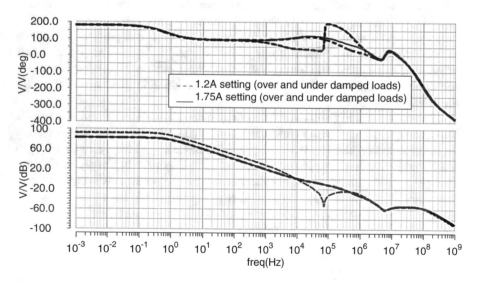

Fig. 4.16 Simulated Bode plot, 1.4 A and 2 A deployment conditions

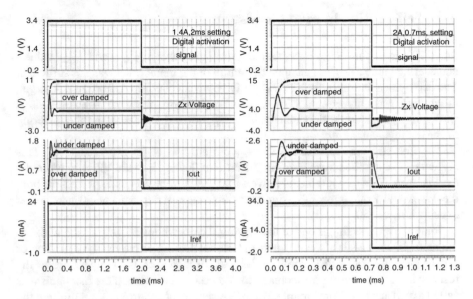

Fig. 4.17 Deployment simulation results for 1.4 A/2 A settings

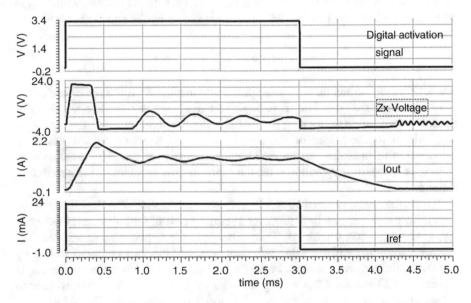

Fig. 4.18 Simulated 1.4 A, 3 ms, 3 mH deployment current

4.6 Self-heating in Metal Resistors

Since 700 μs to 2 ms is an intermediate duration between J_{PEAK} and J_{RMS} values, considering value along to calculate the width of the metal trace was not sufficient. The deployment current was 1.3 A, slightly below level of 1.4 A. For the 2 A target, the current was only around 1.78 A, just barely above the 1.75 A level. This is shown in Fig. 4.19. Additional analysis was performed to understand the lowering of the deployment current.

The first step was to check if the I_{ref} current was at its expected level. This current can be checked by measuring the current flowing into the VZx pin when the load is open. This measurement showed that the current was as expected for both the 1.4 A and 2 A targets. This indicates that the resistor ratio of 60 was not as expected. It was 50.

Joule's law of heating is one of the driving factors while deciding on the width of the metal trace. This is also known as the ohmic heating or the resistive heating or self-heating. As power dissipation in the resistor increases, the temperature of the resistor increases causing an increase to the resistance value. This mechanism is called as self-heating. For example, when 2 A deployment current flows through the resistor r_1 shown in Fig. 4.2, self-heating will cause the resistance r_1 to increase. This means that the 120 mΩ resistor will increase, causing the output current I_{out} to be lower than 2 A. The heating "H" in Joule is shown by the equation below:

$$H \quad \propto \quad I^2R \ t \tag{4.8}$$

where "I" is the current flowing through the conductor, "R" is the resistance of the conductor and "t" is the time. It is important to ensure that the temperature rise should be well within the melting point of aluminium ~ 650 °C. Since the heating is proportional to the square of the current flow, the temperature rise has been measured to be instantaneous on the aluminium metal resistor. The instantaneous heating is approximately within 4 μs, the thermal time constant of the aluminium metal resistor.

Hence, the J_{PEAK} value of 50 mA is not sufficient to allow 2 A flow through the metal. To overcome this limitation, an average value of current density between J_{PEAK} and J_{RMS} has been used. This minimizes the impact due to self-heating. However the sense resistor value will decrease, and this can be compensated by increasing the reference current. In the current design, the reference current I_{ref} is increased to 33 mA for 1.2 A setting and 45 mA for the 1.75 A setting in order to achieve the targets of approximately 1.4 A and 2 A, respectively (Fig. 4.19).

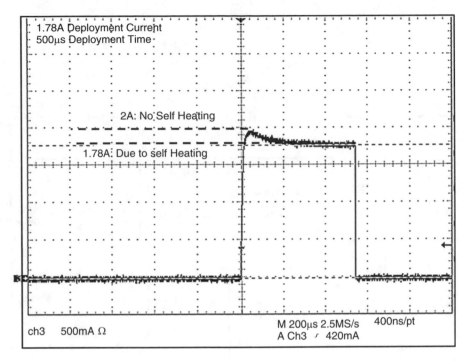

Fig. 4.19 Self-heating of metal resistor during deployment

4.7   Energy Limitation and Inadvertent Current Limitation

Energy limitation is needed in order to prevent the unintended deployment of the
squib. Unintended activation indicates that the current through the HS_FET into the
squib load is higher than 8 mJ. Figure 4.20 shows this scenario. In Fig. 4.21, when
all the HV, MV and LV rails are zero, there is no output current. However if there is
a system fault such that Z_2 gets instantaneously short to ER, a fast transient occurs
on VZ_1, through Deploy unit 2. The C_{gd} Miller capacitor of the HS_FET connected
to VZ_1 gets charged up, and the only way to discharge C_{gd} is to provide a shunt path
to ground.

If there is no discharge path available, the time to discharge C_{gd} can be 6 ms or
higher. This is shown in the simulations in Fig. 4.22. The Zx is at ground, and a $dV/
dt$ of 35 V/300 ns is applied on the VZ_1 pin. The gate of the HS_FET is charged up,
and in the unpowered state of the ASIC, the C_{gd}, C_{gs} of the HS_FET [10] are not
discharged. The current source at the source follower output stage is of very high
impedance, i.e. in the order of 100 MΩ such that the V_{gs} of the HS_FET is not zero.
Hence an additional circuitry is needed to ensure that the C_{gd} and C_{gs} are discharged
such that the V_{gs} of the HS_FET is zero and the peak current through the HS_FET
during the voltage transient is brought towards less than 100 mA (no fire specification) in a short time.

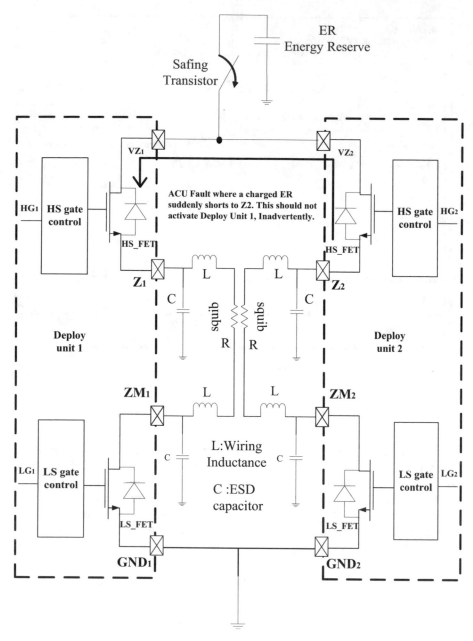

Fig. 4.20 System level fault in unpowered state of ASIC

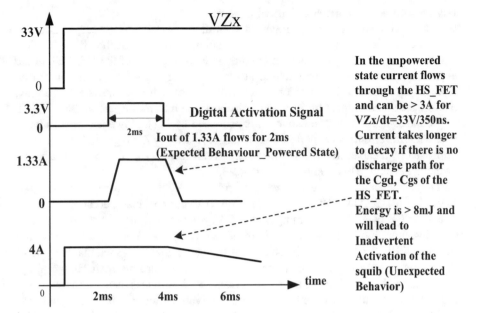

Fig. 4.21 Unpowered state surge current issue

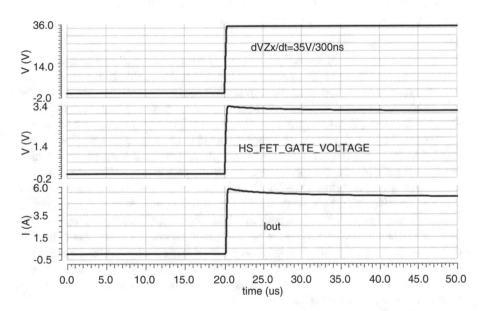

Fig. 4.22 Simulations of the HS_FET in unpowered state during voltage transient without surge current control

High Speed Surge Current Control Circuit

In order to achieve the energy limitation or limit the duration of the surge current, a resistor can be connected between gate and source terminals of the HS_FET. This will ensure that the V_{gs} is less than the threshold voltage V_{th} of the HS_FET in the unpowered state. The current source at the output of the source follower can be replaced by a resistor in the order of 100 KΩ. However for fast transients on VZx, the resistor alone is not sufficient, and a discharge path from the gate of the HS_FET to ground is needed. A good solution is proposed by [11]. However the disadvantage with this circuit is the permanent current flow from VZx to ground. This can be improved by a capacitive feedforward [12] circuit from VZx that powers the active discharge circuit. This is shown in Fig. 4.23.

The capacitive feedforward from VZx powers the rail MVS_SUPPLY_ADSCHG. This ramps from 0 V and gets clamped approximately to 6 V or 7 V even if VZx is ramping to 35 V. Since the digital activation signal is 0 during the unpowered state, the drain of the transistor MD6 is pulled up to the MVS_SUPPLY_ADSCHG rail. This keeps the Ctrl signal high such that the gate of the transistor Mdg is pulled up to ground. This will ensure faster discharge of the gate of the HS_FET, limiting the surge current to 1.5 A in less than 1 μs as shown in Fig. 4.24. In the powered state, the activation signal is high and the MVS_SUPPLY_ADSCHG rail is supplied by the MV rail. This keeps the Ctrl signal 0, such that Mdg is turned *off*. This approach offers the advantage of VZx leakage less than 1 μA in the unpowered state of the ASIC and does not interfere with the current limitation OTA in the powered state.

Measurement Results

The current regulation of the HS_FET is tested with VZx = 33 V. The gate driver Vdd is at 36 V. The input bias current for the driver ASIC was 10 μA from which the reference current was generated. With SPI software, the I_{out} current can be chosen between 1.2 and 1.75 A. The currents were measured to be 1.37 A for 1.2 A setting. Chosen deployment period was 2.1 ms. Similarly the measured current was 1.94 A for the 1.75 A setting. The period was chosen to be 0.75 ms. The current regulation was checked for 1 Ω resistive load, 220 nF load with the inductance being 1 μH and 100 μH. The internal free-wheeling path is verified by observing the Zx voltage going negative to −2.5 V such that the free-wheeling current flows through the HS_FET. This is shown in Fig. 4.25. It also shows the additional measurement results that were performed to check the free-wheeling path by using an inductance of 3 mH in the special mode where the current is regulated to 1.37 A for 2.1 ms. Finally Fig. 4.25 shows the energy limitation on the HS_FET in the unpowered state. This is checked by connecting the Zx pin to ground through a 1 Ω resistor and applying a fast ramp of 35 V in 300 ns on VZx pin. The surge current is around 1 A for 400 ns.

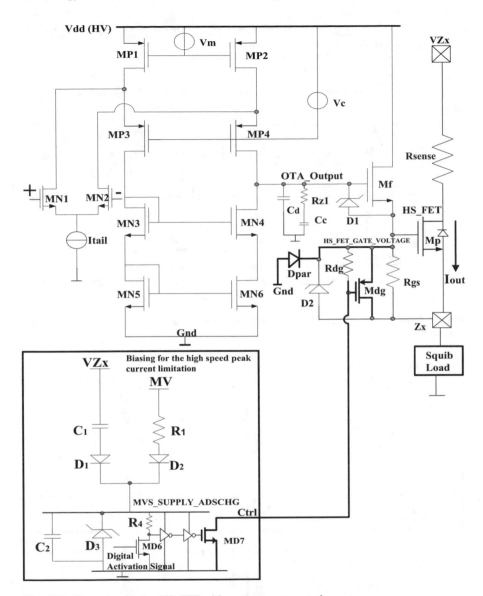

Fig. 4.23 Current regulating HS_FET with surge current control

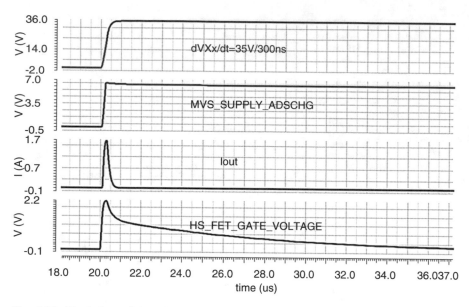

Fig. 4.24 Simulations of the HS_FET in unpowered sate during voltage transient with surge current control

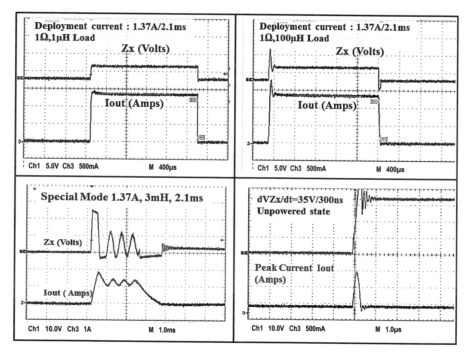

Fig. 4.25 Measured deployment characteristics 1.2 A current setting (1.37 A nominal current) in powered state and surge current control in unpowered state

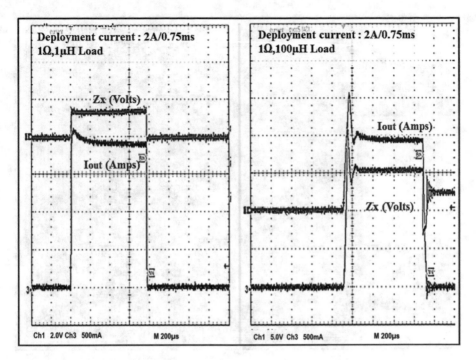

Fig. 4.26 Measured 2 A/0.75 ms deployment current

Figure 4.26 shows the deployment current for the 1.75 A setting (2 A nominal current) for 1 Ω and 1 μH and 100 μH inductive loads.

4.8 HS_FET, Gate Driver Layout

The layout of the HS_FET and its gate driver from a single channel is shown in Fig. 4.27. The layout shows the placement of the HS_FET and metal resistor for current sense. Based on the thermal simulations, the OTA has been placed 400 μm away from the centre of the HS_FET. To the extreme right, the reference current I_{ref} generator module is placed. The HS_FET and the metal resistors have a good copper routing in order to conduct the high currents.

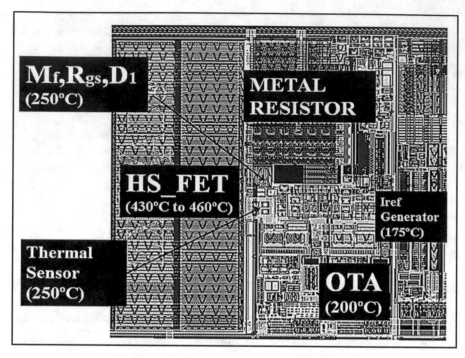

Fig. 4.27 HS_FET and gate driver layout

4.9   Summary and Conclusions

The current regulation method for the HS_FET through a metal sense resistor has been discussed. The need for high open-loop gain of the OTA with the frequency compensation scheme for a wide range of *R-L-C* loads has been presented. The current regulation architecture has an internal freewheeling path provided by the HS_FET itself. This is also verified in a special mode where the deployment time is increased to 4 ms to drive a 3 mH coil. This solution avoids external Zener diodes and is cost-effective. The energy limitation scheme is implemented by using a capacitive feedforward scheme from the drain supply of the HS_FET, and it ensures a faster discharge of the gate capacitances during unpowered state of the driver and does not interfere with the main regulation path of the OTA in the powered state.

References

1. FloTHERM Suite Brochure, Datasheet, *Mentor Graphics*, 2015
2. P.L. Hower, Safe operating area – a new frontier in LDMOS Design, in *Proceedings 14th International Symposium on Power Semiconductor Devices and ICs*, Santa FE, New Mexico, June 2002, pp. 1–8, doi: https://doi.org/10.1109/ISPSD.2002.1016159

3. "Understanding Thermal Dissipation and Design of a Heat Sink," application report, *Texas Instruments*, SLVA462, May 2011
4. T.Y. Wang, C.C.P. Chen, SPICE-compatible thermal simulation with lumped circuit modeling for thermal reliability analysis based on modeling order reduction, *IEEE Computer Society*, 0-7695-2093-6/04, 2004
5. A. Hastings, *The Art of Analog Layout*, 2nd edn. (Pearson International Edition, Prentice Hall, 2001)
6. J. Liening, Interconnect and current density stress – an introduction to electromigration aware design, in *SLIP '05, Proceedings*, pp. 81–88, California
7. S.N. Easwaran, M. Wendt, Current driver circuit, U.S. Patent US 8,093,925B2, 10 Jan 2012
8. B. Razavi, *Design of Analog CMOS Integrated Circuits* (McGraw-Hill, New York, 2001)
9. P. Horowitz, W. Hill, *The Art of Electronics* (Cambridge University Press, Cambridge, 2001)
10. *User Configurable Airbag IC*. L9678, L9687-S, Datasheet, DocID025869, Rev 3.0, STMicroelectronics, May 2014
11. *4 Channel Squib Driver*. E981.17, Datasheet, QM-No. 25DS0073E.00, Elmos, Aug 2011
12. S.N. Easwaran, Driver control apparatus and methods, U.S. Patent US 9,343,898B2, 17 May 2016

Chapter 5
Low-Side Current Regulation and Energy Limitation

The airbag squib drivers have the LS_FET operated in $R_{\text{ds\_on}}$ mode. However to protect the LS_FET against short-to-battery conditions, a current regulation or current limitation is necessary. In this chapter the current regulation of the LS_FET with a senseFET-based current sensing scheme is described. It describes the challenges that are needed to stabilize the LS_FET for the *R-L-C* loads. The same load conditions applied to the HS_FET apply to the LS_FET. The targeted current limit is 3 A for 2 ms. Similar to the HS_FET, the LS_FET needs energy limitation in the unpowered state in order to prevent inadvertent deployment of the squib.

5.1 SenseFET Topology and Current Sensing

In the senseFET current sensing scheme, the senseFET is connected in parallel to the powerFET. In Chap. 2, on the LS current sensing, the current sensing scheme was implemented by connecting the drains of the powerFET and senseFETs in parallel and implementing an OTA for equalizing the source of the FETs. The V_{gs} mismatch [1] will be high between the powerFET and senseFET in this scenario.

It will affect the accuracy of the current limitation or regulation. On top of the V_{gs} mismatch [1], the V_{th} mismatch between powerFET and senseFET is very high if LDMOS FETs are used. Hence it is a better approach to connect the sources of the powerFET and senseFET together in this type of current sensing scheme and use an OTA to regulate the drain voltages. This is used in current mirrors with low V_{ds} operation [2] except that the circuit here is a supplied by HV rail Vdd. This is shown in Fig. 5.1.

© Springer International Publishing AG 2018
S.N. Easwaran, *Current Sensing Techniques and Biasing Methods for Smart Power Drivers*, https://doi.org/10.1007/978-3-319-71982-5_5

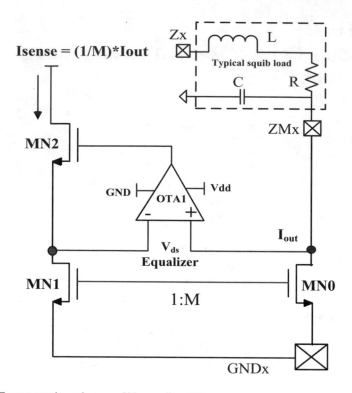

Fig. 5.1   Current sensing schemes – Vds equalizer [4]

For linear mode of operation, I_{sense} is shown in Eq. (5.1).

$$I_{\text{sense}} = \frac{1}{M} \mu C_{\text{ox}} \left(\frac{W_{\text{MN0}}}{L_{\text{MN0}}} \right) (V_{\text{gs}} - V_{\text{th}})\, V_{\text{ds}} \qquad (5.1)$$

For saturation mode of operation, I_{sense} is shown in Eq. (5.2).

$$I_{\text{sense}} = \frac{1}{M} \frac{\mu C_{\text{ox}}}{2} \left(\frac{W_{\text{MN0}}}{L_{\text{MN0}}} \right) (V_{\text{gs}} - V_{\text{th}})^2 (1 + \lambda V_{\text{ds}}) \qquad (5.2)$$

The sensed current is now used for the current regulation loop to control the gate of the powerFET in order to regulate or limit the current during a fault condition.

5.2   Current Limitation Loop

The OTA1 along with the transistors MN0, MN1, MN2, C_{gs}, C_{gd} of MN0 and the R-L-C load form a voltage regulation loop. If there is no OTA1, the MN1, MN2 transistors will not be influenced by the R-L-C load. It is also important to realize

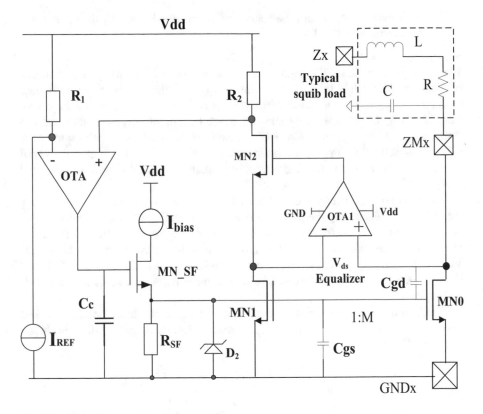

Fig. 5.2 Low-side current regulation topology

that the LS_FET MN0 along with the *R-L-C* load forms a gain stage unlike the HS_FET where the output stage was an attenuation stage. Hence the gain at the current limitation stage can be less, and a 60–75 dB gain is sufficient to achieve the desired accuracy.

$$I_{\text{limit}} \approx \left(\frac{R_1}{R_2}\right) MI_{\text{ref}}$$

$$\Rightarrow I_{\text{limit}} \approx \left(\frac{312\ K\Omega}{225\ \Omega}\right)(64)(34\,\mu A) \sim 3A$$

(5.3)

The drain of MN1 is regulated by the OTA1 to the drain voltage of MN0. In the improved topology presented here in Fig. 5.2, the OTA1 equalizes the V_{ds} of MN0 and MN1 in order to sense the current accurately. When the current through MN0 increases, the sensed current MN1 will increase. When the output of MN2 is regulated to a reference voltage generated by the resistor R_2 and current I_{ref} by using another OTA as shown in Fig. 5.2, the current through the senseFET is regulated, and thereby the current through the powerFET MN0 is limited or regulated. A source follower MN_SF is used to isolate the OTA from the large gate capacitance of MN0, and the resistor R_{SF} ensures $V_{gs} < V_{th}$ to prevent any inadvertent turn *on* of MN0.

Small-Signal Analysis

We know from the HS_FET regulation that a resistor between the gate and source of the powerFET might be critical for the energy limitation in the unpowered state to keep the $V_{gs} = 0$ and avoid any inadvertent turn on. The challenge is to ensure that the OTA, OTA1 are stable and the overall loop stability is maintained. To stabilize the OTA loop, capacitance C_c would have to be added at the output of the OTA. This will create a pole, and since C_{gd}, C_{gs} of the LS_FET is high, the overall stability of the loop for the high inductive loads is checked from small-signal analysis.

The C_{gd} and C_{gs} of the powerFET are assumed to be equal to C_a for simplifying the analysis. Considering the powerFET and the inductive load, along with C_{gd} of the powerFET, the transfer function of this power stage from its gate to the drain is already of the second order. It is shown in the equation below. So the mathematical analysis for stability of the loop is performed without OTA1 and the source follower stage in order to get an understanding of the main poles and zeros at the R-L-C output stage.

For L in µH and for $r_0 > 2\ \Omega$, with g_{mp} and g_{dp} as the transconductance and output conductance of MN0, the transfer function $H_p(s)$ from gate to drain is shown in Eq. (5.4).

$$H_p(s) \sim -\ \frac{\frac{g_{mp}}{g_{dp}}\left[1 + \left(\frac{sL}{R} - \frac{sC_a}{g_{mp}}\right)\right]}{\left[1 + \frac{sL}{R} + s^2\frac{LC_a}{g_{dp}}\right]} \qquad (5.4)$$

Since the power stage itself is a second-order equation and in order to stabilize the current regulation loop for a wide range of inductive loads, the OTA should be a single-stage output [3]. Hence the OTA is designed with a current-based input stage unlike the HS_FET current regulation loop. This is shown in Fig. 5.3. The MP1, MP2 input stage of the OTA is a common gate or a current input stage.

The small-signal equivalent is shown in Fig. 5.4 from which the transfer function V_o/V_a can be deduced.

The C_{gs} of the powerFET is sufficient to stabilize the OTA. The transfer function for the small-signal model is shown below. The transfer function has two poles and one zero. This gives a good indication that the system is stable theoretically with a good phase margin. This is shown in Bode plot in Fig. 5.5.

For L in µH, $r_0 > 1\ \Omega$, and $g_{mp} = g_{dp}$, the open-loop DC gain, poles, zero and the unity gain frequency are shown below. The powerFET, senseFET transconductance, load, current sense resistor connected to senseFET are fixed and only the OTA parameters can be changed. This is shown in Table 5.1.

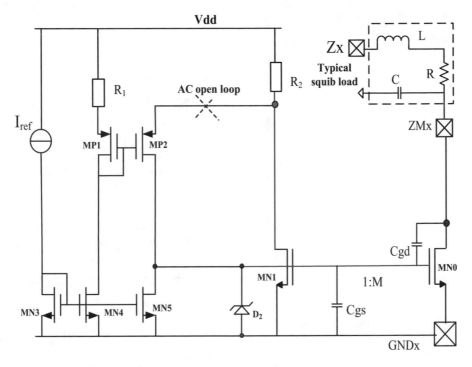

Fig. 5.3 Low-side current regulation circuit for small-signal analysis

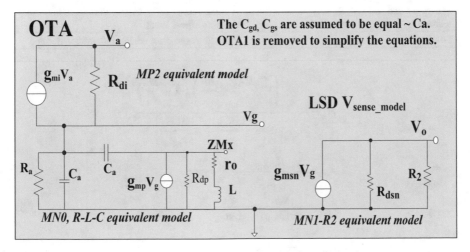

Fig. 5.4 Small-signal model for LS_FET current regulation loop

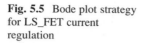

Fig. 5.5 Bode plot strategy for LS_FET current regulation

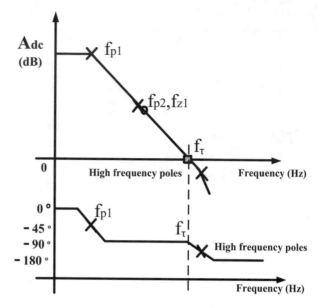

Table 5.1 Design parameters and choices

Amplifier stages	Parameter	Degrees of freedom	Condition
OTA	$\frac{g_{mi}g_{msn}R_2}{g_{di}}$	√	Open-loop gain, stability
PowerFET	g_{mp}	×	R_{ds_on}, thermal
SenseFET and resistor	$g_{msn}R_2$	×	Head room (18–V supply variation) and minimum signal level
Load	r_o, L	×	Squib resistance, wiring inductance defined

$$H(s) \sim -\frac{\left(\dfrac{g_{mi}R_{di}g_{msn}R_2}{g_{dp}}\right)\left(1+\dfrac{sL}{r_o}\right)}{\left[1+\dfrac{3C_a}{g_{di}}+s^2\dfrac{3LC_a}{r_o g_{di}}\right]}$$

$$A_{dc} = \frac{g_{mi}R_{di}g_{msn}R_2}{g_{dp}}$$

$$f_{p1} = \frac{g_{di}}{2\pi(3C_a)} = \frac{g_{di}}{6\pi C_a}$$ (5.5)

$$f_{p2} = \frac{r_o}{2\pi L}$$

$$f_{z1} = \frac{r_o}{2\pi L}$$

$$f_\tau = \frac{g_{mi}g_{msn}R_2}{6\pi C_a}$$

With these equations, the open-loop DC gain, poles, zero and the unity gain frequency are calculated as shown below. Assuming the following parameters

$$g_{mp} = g_{dp} = 2.2S, g_{mi} = 100\mu S, g_{msm} = 10mS, R_{di} = 1G\Omega \, (g_{di} = 1nS), R_2$$
$$= 150\Omega, C_a = 50pF$$

the loop gain, poles, zero and the unity gain frequency are shown below.

$$H(s) \sim - \frac{\left(\dfrac{g_{mi}R_{di}g_{msn}R_2}{g_{dp}}\right)\left[1 + \dfrac{sL}{r_o}\right]}{\left[1 + s\dfrac{3C_a}{g_{di}} + s^2\left(\dfrac{L}{r_o}\dfrac{3C_a}{g_{di}}\right)\right]}$$

$$A_{dc} = \frac{g_{mi}R_{di}g_{msn}R_2}{g_{dp}} = \frac{(100.10^{-6})(1.10^{+9})(10.10^{-3})150}{2.2} \sim 68181$$

$$\Rightarrow A_{dc} = 20\log_{10}(68181) = 20(4.83) \sim 97 \text{ dB}$$

$$f_{p1} = \frac{g_{di}}{2\pi(3C_a)} = \frac{1.10^{-9}}{6\pi(50.10^{-12})} = 1.06 \text{ Hz}$$

$$f_{p2} = \frac{r_o}{2\pi L} = \frac{8}{6.282(100.10^{-6})} \sim 12.8 \text{ kHz}$$

$$f_{z1} = \frac{r_o}{2\pi L} = \frac{8}{6.282(100.10^{-6})} \sim 12.8 \text{ kHz}$$

$$f_{\tau} = \frac{g_{mi}g_{msn}R_2}{2\pi(3C_a)} = \frac{(100.10^{-6})(10.10^{-3})150}{6(3.141)(50.10^{-12})} = 159 \text{ kHz}$$

Phase Margin calculation

$$\Phi_m = 180° - \arctan\left(\frac{f_{\tau}}{f_{p1}}\right) - \arctan\left(\frac{f_{\tau}}{f_{p1}}\right) + \arctan\left(\frac{f_{\tau}}{f_{z1}}\right)$$

$$\Rightarrow \Phi_m = 180° - \arctan\left(\frac{159K}{1.06}\right) - \arctan\left(\frac{159K}{12.8K}\right) + \arctan\left(\frac{159K}{12.8K}\right) = 90°$$

The Bode plot for underdamped and overdamped loads is shown in Fig. 5.6a, b, respectively.

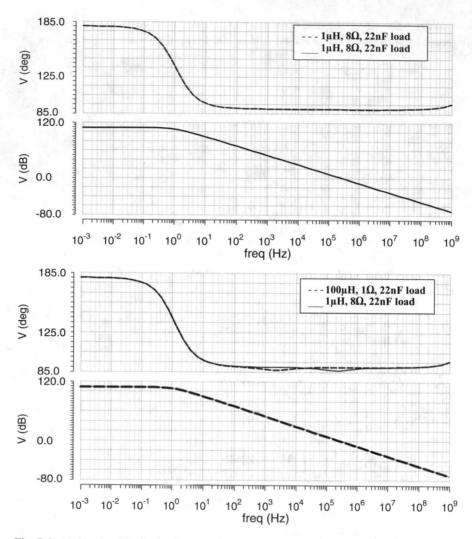

Fig. 5.6 (**a**) Simulated Bode plot for the 8 Ω load. (**b**) Bode plot for the model for 1 Ω load

This makes the loop, a single-pole system, and is independent of the inductance. In other words the phase margin of the overall loop by just relying on the C_{gs} and C_{gd} of the powerFET is 90° indicating very good stability.

Deviations from the Small-Signal Model
The design should focus towards stability but also in ensuring that the transistors are operated within their SOA. If low-/medium-voltage (LV/MV) components are used to achieve excellent matching, appropriate design measures have to be taken to

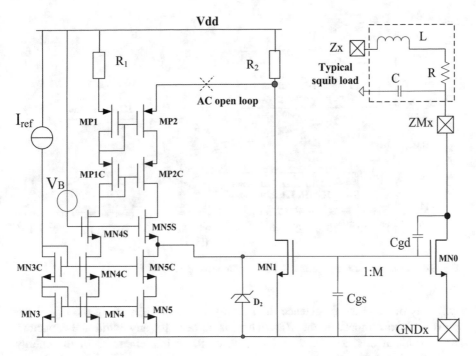

Fig. 5.7 Low-side current sensing scheme

ensure that the $|V_{ds}|$ and $|V_{gs}|$ of the LV and MV transistors are within the allowed SOA. For instance, the $|V_{ds}|$ of MV components cannot exceed 6 V, and the $|V_{gs}|$ is within 12 V. On the HV component, the $|V_{ds}|$ can be as high as 40 V, whereas the $|V_{gs}|$ is limited to 12 V. These constraints lead to the addition of cascode transistors, which will alter the assumed values of transconductance g_m and output conductance g_{ds}. The circuit topology comprising of the cascode transistors for high-voltage protection is shown in Fig. 5.7. The input-stage transistors MP1 and MP2 are MV transistors. They are cascaded with HV transistors MP1C and MP2C to ensure that the $|V_{ds}|$ of MP1, MP2 is within 6 V. The current mirrors MN4 and MN5 are cascoded with MN4C and MN5C in order to increase the output impedance. These cascaded transistors are medium-voltage (MV) transistors. Since the $|V_{gs}|$ of the powerFET cannot exceed 12 V, additional shielding transistors MN4S and MN5S are added. The gate voltage of these transistors does not exceed more than 12 V such the gate voltages of MN0, MN1 are limited to 12 V. The MN5S transistor operates in the linear region when the current is regulated.

The Bode plot for the low-side current sensing scheme is shown below. The deviations in the transconductance and the output impedance will reduce the phase margin as shown in Fig. 5.8 for the underdamped load conditions like the 1 Ω, 100 μH R-L combination. However the minimum phase margin ϕ_m obtained under this condition is 20° making the system stable. Due to the ϕ_m degradation, it is wise to avoid the source follower stage as the additional pole from the source follower

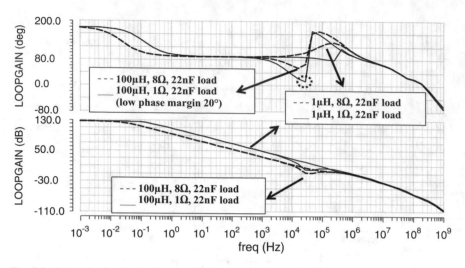

Fig. 5.8 Bode plot for the implemented LS current sensing scheme

will only degrade the ϕ_m. Hence the stability is achieved by using the C_{gs} of the powerFET to compensate the OTA. There is no need for any additional capacitor. As a result of this, a different scheme needs to be implemented for the energy limitation in the unpowered state.

Additional analysis showed that the ESD capacitor on the ZMx pin helps to improve the phase margin. This will introduce a zero in the transfer function such that the phase margin improves. The capacitor range between 22 nF and 220 nF will increase ϕ_m only by 5°, and in order to gain additional phase margin, the capacitor on the drain ZMx of the LS_FET should be in the order of 2.2–4.7 µF. This is shown in Fig. 5.9.

5.3 V_{ds} Equalizer (OTA1) in the Current Regulation Loop

The V_{ds} equalizer is now brought into the stability of the loop. The OTA1 has to be designed such that the poles from OTA1 are outside the f_τ. This will ensure that the stability of the overall loop is unaffected. The schematic [4] of the single-stage OTA1 is shown in Fig. 5.10. The input transistor is a common gate stage or a current input amplifier. The gate of the input stage is biased through a diode-connected transistor, MN11. The drain of MN12 is the output of the OTA1, and it drives the cascode FET of the current sensing transistor MN1. The gain of the OTA1 is in the range of 20 dB, and the overall loop formed by OTA1 along with MN0, MN1 and MN2 provides around 65 dB gain with a phase margin of 70°. The gain of OTA1 is shown below in the Eq. (5.6).

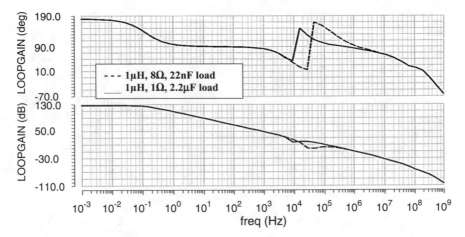

Fig. 5.9 Bode plot for the LS current sensing scheme with 22 nF and 2.2 μF capacitors on ZMx pin

Fig. 5.10 OTA1 schematic, common gate amplifier [4]

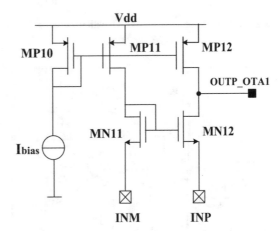

$$A_{dc} = \frac{g_{m2}}{g_{dn2} + g_{dp2}}$$

$$g_{m2} = 1\mu S \quad \text{and} \quad \frac{1}{g_{dn2} + g_{dp2}} = 10M\Omega \qquad (5.6)$$

$$\Rightarrow A_{dc} = \left(10^{-6}\right)\left(10^{-6}\right) = 10 = 20\log_{10}(10) = 20 \text{ dB}$$

The Bode plot response of the current regulating LS_FET with small signal variation on OTA1 is shown in Fig. 5.11. The OTA does not have any small signal variation in this case. The OTA1 loop is stable. Overall both the OTA1 and OTA loops have been proven to be individually stable. If every loop is stable, the overall loop is stable and can be verified by transient simulations.

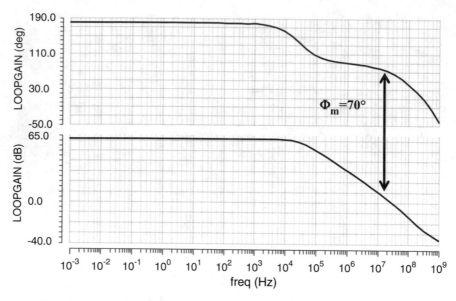

Fig. 5.11 Bode plot for the OTA1 with no small-signal variation on OTA

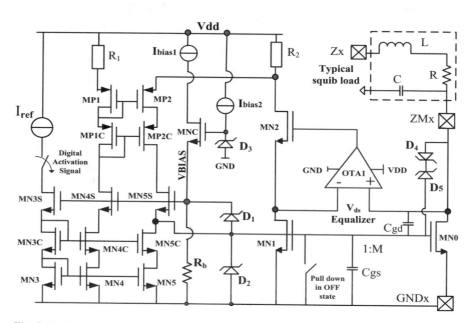

Fig. 5.12 Schematic of the LS driver

The overall topology of the LS_FET current limitation is shown in Fig. 5.12. The additional biasing network formed with I_{bias1}, MNC and R_b biases the shielding transistor MN3S, MN4S and MN5S. The gate of the transistor MNC is biased

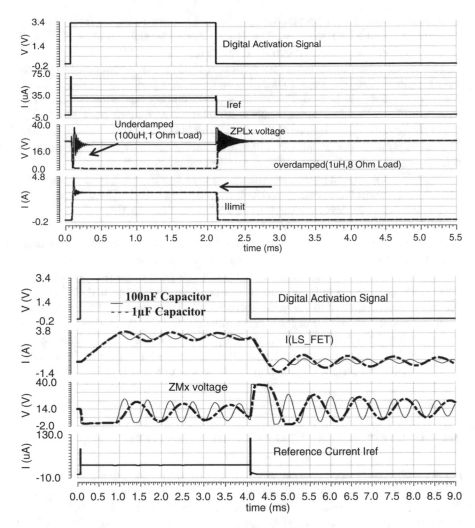

Fig. 5.13 (**a**) Transient response of the LS driver. (**b**) Transient response for LS driver with 3 mH and 22 nF/1 μF capacitor

through the current source I_{bias} from Vdd. The transistor MNC width and the resistor R_b are chosen such that VBIAS would not exceed 12 V. This helps in ensuring that the transistors MN3-MN3C, MN4-MN4C and MN5-MN5C are built with MV components to achieve excellent matching. MN3S, MN4S and MN5S transistors are HV transistors whose V_{ds} rating is 40 V. The Zener diodes D_1 and D_2 are used to protect the V_{gs} of the FETs MN5S and MN0, MN1. These two diodes are reversely biased in the normal current regulation mode. D_4 and D_5 Zener diodes are gate-drain clamps used to limit the voltage on the ZMx pin during the free-wheeling condition when the FET turns *off*. The D_4–D_5 diodes breakdown after 35 V, and they conduct the free-wheeling current through the LS_FET MN0.

The transient simulation is set up such that a 25 V supply is connected to the Zx pin of the device and the LS_FET MN0 is enabled for a period of 2.3 ms. A 22 nF capacitor on ZMx pin is used. The stability of the current limitation is verified as shown in Fig. 5.13a. The current limit is represented as I_{limit} or I(LS_FET). It is stable for both the underdamped (1 Ω, 100 μH) and (1 Ω, 1 μH) overdamped loads. The underdamped load condition will show some overshoot and slight ringing behaviour prior to settling to the 3 A regulation value.

At 35 V Zx level, the power dissipation on the LS_FET for 2.3 ms is ~105 W, and the over-temperature protection circuit will deactivate the LS_FET prematurely after 1 ms. Hence Zx level is chosen to be 25 V instead of 35 V.

An option in the squib driver ASIC was available to extend the activation time to 4 ms. This is intended to drive inductive loads in mH range. The SCB fault condition is tested at 14 V instead of 25 V since the duration is 4 ms. This reduces the power dissipation in the LS_FET else too much power dissipation will cause over-temperature protection and will prematurely deactivate the LS_FET. Wider activation time is needed to allow the current build up in the coil. For such larger coils, the phase margin will degrade and influence stability. Analysis showed that by increasing the external capacitor, the loop can be made stable. Simulations in Fig. 5.13b show heavy ringing with a smaller capacitor of 100 nF, and the ringing gets lower, and the current limit stability improves with a bigger capacitor of 1 μF.

5.4   Unpowered State Energy Limitation Circuit

Energy limitation is needed in order to prevent the unintended deployment of the squib. Unintended activation indicates that the current through squib flowing into the LS_FET is higher than 8 mJ. In Fig. 5.14, when all the HV, MV, LV rails are zero, there is no output current. However if a SCB condition occurs on the Zx pin, there will be a faster transient on the ZMx pin. The Miller capacitor C_{gd} of the LS_FET connected to ZMx pin gets charged up, and a way to discharge the C_{gd} is to provide a shunt path to ground.

From this architecture it can be seen that the gate voltage of the powerFET MN0 starts to increase due to an instantaneous voltage raise on ZMx by coupling through the C_{gd} capacitance. As the gate voltage starts to raise, the diode D_1 gets forward biased and will discharge the C_{gd} capacitor through the resistor R_b. This resistor is typically in the order of 100 KΩ. This leads to a RC time constant of 5 μs indicating that the steady-state current of lower than 100 mA would be approximately 2–3RC, i.e. 10–15 μs. Theoretically if the resistor R_b is chosen to be smaller, then there is no need for an additional discharge path. However the design constraints from the current limitation loop will not allow R_b to be lower.

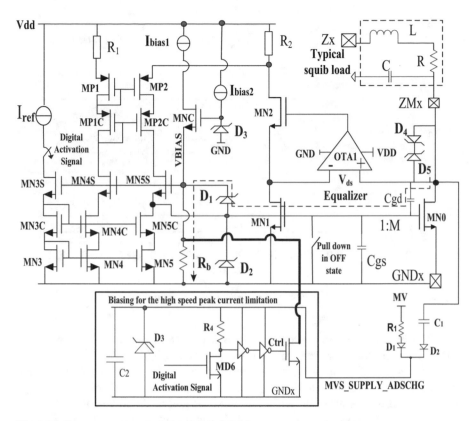

Fig. 5.14 Surge current control circuitry in LS driver

In order to achieve the desired VBIAS, the bias current I_{bias1} has to be increased in the order of mA to achieve less than 5 μs. This discharge is present intrinsically in this current limitation loop [5].

However if in some process technologies the C_{gd} ends up on the higher side, for instance, 100 pF, then the intrinsic discharge path will be extremely slower. In order to speed up the discharge time, a surge current control circuit similar to the surge current control circuit for the HS_FET is added. This is shown in Fig. 5.14. A dV/dt of 35 V/300 ns is applied on the Zx pin, and the current flowing through the LS_FET is measured to check for the peak current and the discharge time. Lower inductance is used for these tests as the peak current is higher with lower inductance.

The surge current can be improved by the high-speed current limitation method similar to the HS_FET high-speed surge current control [6]. A feedforward capacitor from ZMx is used to generate the supply when there is a voltage transient on the ZMx pin. During the powered state, MV rail is available and MVS_SUPPLY_ADSCHG is MV-0.7 V. The gate of the MD6 transistor will be

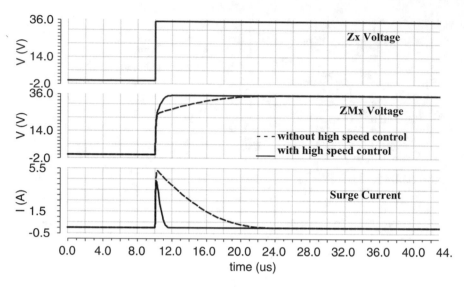

Fig. 5.15 Simulations showing the high-speed surge current control

at 0 V when there is no deployment. This will ensure that the Ctrl signal at ~ MV-0.7 V will pull down the gate of MN0 to ground (GNDx).

As inferred from simulation results shown in Fig. 5.15, the intrinsic or the slow discharge path takes 20 μs to discharge the C_{gd} in the unpowered state. The energy in this period is way above the 115 μJ no-fire spec and could potentially be close to the all-fire energy specification of 8 mJ. This could deploy the squib inadvertently. The high-speed surge current control circuit for faster discharge discharges the Miller capacitor C_{gd} in 1.5 μs.

5.5   Measurement Results

The current regulation of the LS_FET is tested with Zx = 25 V. The gate driver Vdd is at 33 V. The input bias current for the driver ASIC was 10 μA from which the reference current of 30 μA was generated. The deployment period was set to 2.3 ms and the measured current was 3 A. The current regulation was checked for 1 Ω resistive load, 22 nF load with the inductance being 1 μH (overdamped case) and 100 μH (underdamped case). This is shown in Fig. 5.16. The 3 mH inductive load was also tested with 22 nF and 1 μF capacitor in order to show the improved transient response by the choice of external capacitors. The surge current tested by applying 35 V in 300 ns is around 3 A for 1.5 μs. Both measurements are shown in Fig. 5.17.

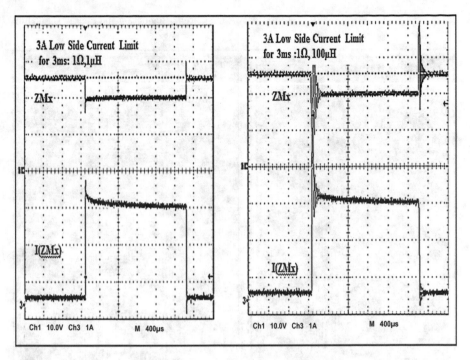

Fig. 5.16 LS driver current limitation for under- and overdamped loads

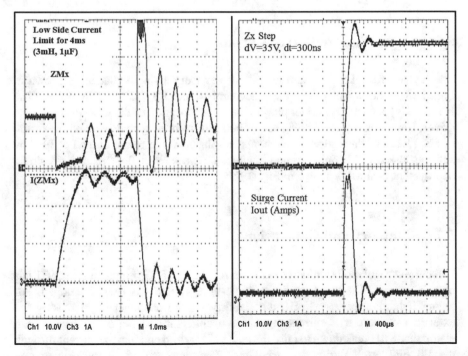

Fig. 5.17 LS driver current limitation for 3 mH, 1 Ω, 1 μF *R-L-C* load and surge current measurement during unpowered state

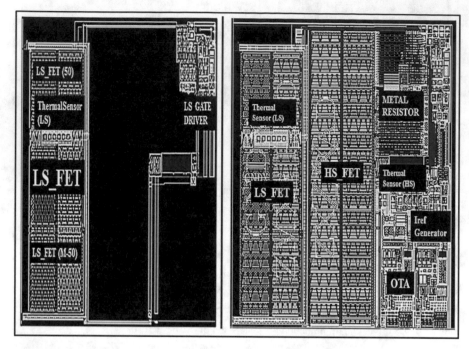

Fig. 5.18 (**a**) LS_FET and its gate driver. (**b**) Single deployment channel with HS_FET, LS_FET and their gate drivers

5.6 LS_FET, Gate Driver Layout

The layout of the LS_FET and its gate driver from a single channel is shown in Fig. 5.18a. The layout shows the placement of the LS_FET and its driver. The HS_FET layout and gate driver placement set the location of the low-side (LS) gate driver. Here it ended up at a distance of 900 μm from the gate driver and the centre of LS_FET. The thermal sense elements are placed at the centre of the LS_FET by splitting the LS_FET into parallel segments. The overall floor plan for a single deployment channel is shown in Fig. 5.18b.

5.7 Summary and Conclusions

The current regulation (limitation) method for the LS_FET through a senseFET has been discussed. The topology for a single-pole system has been developed. For higher inductive loads ~ 3 mH, the stability can be achieved by increasing the external capacitor. The current regulation architecture has an internal gate-drain clamp that will break down to raise the gate voltage and conduct the current through the LS_FET itself during free-wheeling. The presence of an intrinsic energy

limitation scheme is discussed. In order to further speed up the surge current control, a capacitive feedforward scheme from the drain supply of the LS_FET is implemented. The concept is similar to surge current control of the HS_FET,and it ensures a faster discharge of the gate capacitances during unpowered state of the driver and does not interfere with the main regulation path of the OTA in the powered state.

References

1. M.J.M. Pelgrom, A.A.D.C.J. Duinmaijer, A.P.G. Welbers, Matching properties of MOS transistors. IEEE J. Solid State Circuits **24**(5), 1433–1440 (1989)
2. B. Razavi, *Design of Analog CMOS Integrated Circuits* (McGraw-Hill, New York, 2001)
3. K. Laker, W. Sansen, *Design of Analog Integrated Circuits and Systems* (McGraw-Hill, New York, 1994)
4. V. Ivanov, I. Filanovsky, *Operational Amplifier, Speed and Accuracy Improvement* (Kluwer, Boston, 2004)
5. S.N. Easwaran, Electronic device for controlling a current, U.S. Patent US 8,553,388B2, 8 Oct 2013
6. S.N. Easwaran, Driver control apparatus and methods, U.S. Patent US 9,343,898B2, 17 May 2016

Chapter 6
Biasing Schemes and Diagnostics Circuit

The airbag squib drivers operate with multiple supply voltage (MSV) levels, viz. the LV, MV and HV rails. Our goal is to ensure that there is no inadvertent activation of the drivers that will lead to the activation of the HS_FET and LS_FET. In addition to this inadvertent activation, diagnostic circuits for measuring the squib resistance are required. In this chapter, a robust biasing scheme that prevents inadvertent activation of the powerFETs, high currents triggered by tri-stated circuits, too high output voltage on the output buffer with a voltage and current selector biasing scheme is presented. In addition to the biasing scheme, the design of a bidirectional fault-tolerant current limited voltage source (CLVS) that is mandatory for the squib resistance measurement is presented. Bidirectional fault-tolerant CLVS indicates both short to battery and short to ground and can be present simultaneously.

6.1 Fault Mode Supply Generator and Current Selector Circuits

In Chaps. 4 and 5, the current regulation loops for both the HS_FET and the LS_FET were analysed, and a topology that regulates the current in the powered state and limits the energy due to dynamic faults in the unpowered state has been implemented. These current regulation loops are activated only when needed and turned *off* otherwise by the activation signal (LV domain) from the digital core. We have to ensure that this control signal is in the correct state after getting it level shifted to either MV or HV rail. This has to be ensured under any phase of the power up. In addition an input bias current is required during any phase of operation in order to avoid the tri-state situation for comparators, HV level shifters.

As the LV rail is available only during phase III operation, the need for critical circuits that are independent of the LV rail is mandatory for all three phases of

© Springer International Publishing AG 2018
S.N. Easwaran, *Current Sensing Techniques and Biasing Methods for Smart Power Drivers*, https://doi.org/10.1007/978-3-319-71982-5_6

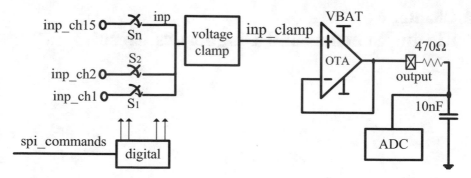

Fig. 6.1 Output buffer and diagnostic switches multiplexing different channels

operation. The existing solutions [1–4] have disadvantages, and so a biasing scheme
built up with the combination of the voltage selector and maximum current selector
[5] is implemented.

At Phase I and Phase II power up situations, the main circuits like the bandgap
references, bias current generators, oscillator and some voltage monitoring circuits
have to be enabled. In the considered system, the serial peripheral interface (SPI)
commands from the microcontroller are processed by the digital logic to activate
switches $S_1, S_2 \ldots S_n$, to measure voltages across different channels by means of an
ADC, as shown in Fig. 6.1. The additional goal is to ensure that the output buffer
voltage level during all three phases of operation does not exceed 5.5 V. These
diagnostic circuits are enabled only during the phase III of operation. However
critical voltage monitoring circuits have to be permanently active to detect the
presence of fault that could be present in the system during power up. Due to the
enabling of some of these modules during the initial power up phase, the typical
quiescent current on the HV rail is specified to be 4 mA.

Fault Mode Voltage Supply (FMVS)

Fault mode voltage supply is based on the maximum voltage selector concept. The
intention behind this is to generate a supply that provides the current during phase I.
Once MV rail is available during phase II, the maximum load current is supplied by
the MV path. This FMVS circuit generates a maximum level based on the voltage
clamped HV stage and the MV rail. This is shown in Fig. 6.2. If only the HV supply
is used, the power dissipation across the transistor is very high and needs to be very
big in area to provide the current load. Hence dominant and non-dominant current
paths are required to optimize the power dissipation. MAVS output at phase I will
be set at 3.8 V. MN2 is made stronger than MN1 in order to set MV as the dominant
supply during phase II and III operation.

During phase I, HV is the dominant supply. The gate of MN1 is clamped by an
on-chip Zener diode to ensure that the MAVS rail is below 6 V. MAVS output at
phase I will be set at 3.8 V. MN2 is dimensioned stronger than MN1 to ensure MV
as the dominant supply during phase II and III operation.

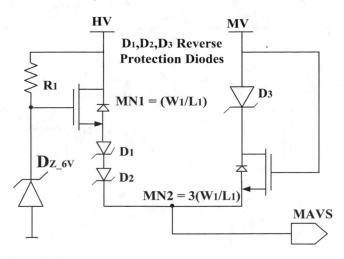

Fig. 6.2 Fault mode voltage supply

Fig. 6.3 Cascaded level
shifter with common source
stage

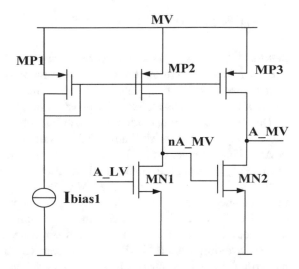

During phase I and II where the LV rail is not available, the level shifter should
still provide the correct output. As the cross coupled level shifters can provide
wrong outputs due to the absence of the LV rail, these level shifters would need a
resistor to ground at its output to resolve the tri-stated situation. This resistor should
be in the order of 100 KΩ to 1 MΩ. This will increase the current consumption of
each level shifter and so is not a preferred choice. Hence a cascaded inverting,
common source stage is the preferred way to level shift the signals from LV to MV
rail. This is shown in Fig. 6.3.

However the bias current I_{bias1} should be present during any phase of operation.
During phase I operation, the I_{bias1} can be a smaller current say 200 nA to ensure

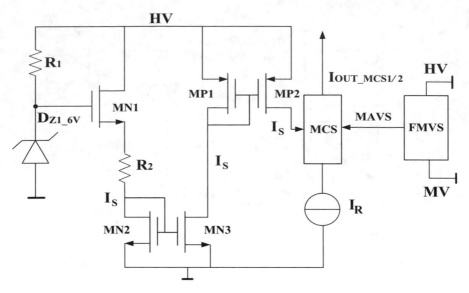

Fig. 6.4 Biasing scheme with FMVS and MCS

that the outputs are defined in the correct way. As the driver ASIC enters phase II operation, more circuits start to operate, and the on-chip junction temperature will start to rise, and the temperatures of MN1 and MN2 will rise even though the ambient temperature is the same for phase I and phase II.

During the phase II operation, I_{bias1} can be increased to 1 μA so that the leakage current of the transistors MN1 and MN2 does not influence the outputs nA_MV and A_MV. For example, if the high-temperature leakage on MN1 is 100 nA and the mirrored I_{bias1} current through MP2 is at 200 nA, the output nA_MV will be at a value much lower than being at MV level. To avoid this situation, it is recommended to increase the I_{bias1} to 1 μA once the ASIC enters the phase II operation. This is achieved by using the maximum current selector. This approach introduces a gradual change in the I_{bias1} from a lower current to a higher current, thereby avoiding situations where this current can momentarily drop to 0 or will have oscillations if the change is based on a comparator decision-based switching. This circuit was discussed in Chap. 3 explicitly, and the overall biasing scheme will look like as shown in Fig. 6.4.

Start-Up and Reference Current

When the HV rail is available during phase I, the start-up current I_S is generated through MN1, R_2 and MN2. The gate of the transistor MN1 is clamped to 6 V by a Zener diode. This current is typically 2 μA. The main reference current I_R is 0 during phase I. When the MV rail is available, the I_R is generated and is about 10 μA. This is a precise current generated from bandgap and is precise with less temperature drift. During phase II operation, the start-up current I_S is still 2 μA. These two currents are fed into the maximum current selector (MCS) circuit. The MCS circuit

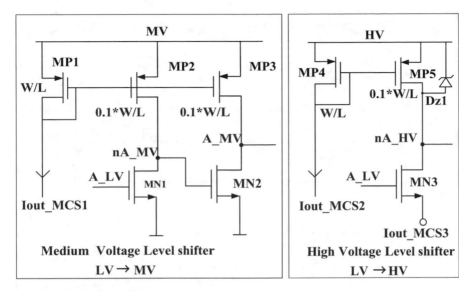

Fig. 6.5 Common source level shifters

is supplied by the output of the FMVS circuit. As we have seen in Chap. 3, the MCS circuit outputs the maximum of the two input currents.

During phase I, the output current $I_{\text{out\_MCSx}}$ is the maximum of the currents 2 µA and 0, i.e. 2 µA. During phase II operation, the output current $I_{\text{out\_MCSx}}$ is the maximum of the currents 2 µA and 10 µA, i.e. 10 µA. The value of x is $\{1, 2, \ldots, n\}$, indicating that several current outputs can be generated by the NMOS current mirrors MN6, MN7, etc. These outputs are fed to the common source level shifter and to the LV to HV level shifter as shown in Fig. 6.5, thereby enabling a level shifter independent of the LV supply. The transistors MP2, MP3 and MP5 are 10× smaller than MP1 and MP4 so that 200 nA is used during phase I and 1 µA during phase II as intended. Additional current outputs of the MCS circuit is used to bias the comparators, output buffer, etc.

The simulation results for this biasing scheme are shown in Fig. 6.6. The start-up current I_S is 2 µA, and the reference current I_R is 10 µA. MAVS node is at 3.65 V. Slight dip on MASV node happens prior as MV rail ramps up. The output current $I_{\text{out\_MCS1}}$ gradually changes from 2 to 10 µA when the MV rail is present and vice versa when MV rail is not present. The simulated quiescent current at the complete chip level was 4 mA.

6.2 Voltage Clamping on the Output Buffer

The input to the operational transconductance amplifier (OTA) functioning as an output buffer ranges from 100 mV to 25 V, whereas the output voltage should be in the low-voltage range ~ 5.5 V. This ensures that the analog to digital converters

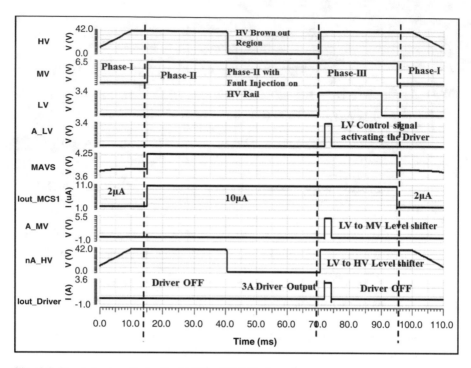

Fig. 6.6 Simulation results for the FMVS – MCS biasing scheme

(ADCs) connected to the high-voltage (HV) OTA do not face destruction due to overvoltage. ADCs are designed with low-voltage (LV) components for better matching and precision. Therefore, voltage clamping of the OTA needs to be guaranteed from phase I to III. During phase I, the supply to the OTA would be ramping up, and the input to the OTA could be floating such that its output can exceed 5.5 V. The existing solutions of using a Zener-based clamping have a wide spread in the output level due to the variation of Zener breakdown from 5.7 to 8 V. This is shown in Fig. 6.7a. Most of the process technologies do not offer precise Zener diodes, and the breakdown voltage can drift over temperature and current density. 8 V is too high for the ADC loads. Here the output voltage is clamped by adding a regulation loop through OTA1 at the output. This is shown in Fig. 6.7b. where the class AB output buffer [2] has another path in parallel when inputs exceed vclamp level. The disadvantage with this approach is the M1 leakage at on-chip junction temperatures (~175 °C) [3] that reduces the output impedance of the OTA, thereby increasing the offset. This is a significant concern for the inputs in the ADC signal range.

Figure 6.8 shows an improved technique by which the output voltage can be clamped precisely [6]. A medium-voltage (~8 V to 10 V)-rated component-based comparator, multiplexer (mux) and the OTA are used for the clamping. It has two advantages: The clamping voltage can be chosen independent of the process

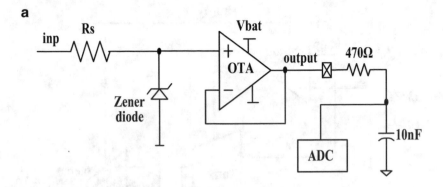

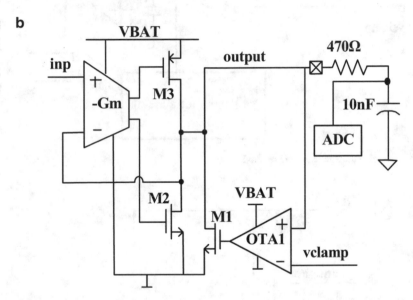

Fig. 6.7 (**a**) Existing solution – Zener diode-based voltage clamping. (**b**) Existing solution – OTA1-regulated voltage clamping method

parameters, and the clamp level is precise and is determined purely by design parameters, like matching of the devices, etc. It also ensures high output impedance as there is no additional leakage such as in Fig. 6.7b. The comparator uses the Zener clamp D_3 to protect its input transistors from exceeding 8 V.

The "supply" and "vclamp" are generated from an internal NMOS regulator supplied from the battery V_{BAT} as shown in Fig. 6.9. LDMOS transistor is used as the powerFET in this regulator design. The bandgap voltage is used as the reference voltage. The stable bias current for the low-dropout regulator (LDO) is generated by using the current selector and fault mode supply in a multiple supply voltage design. Miller compensation by using the capacitor C_m stabilizes the regulator.

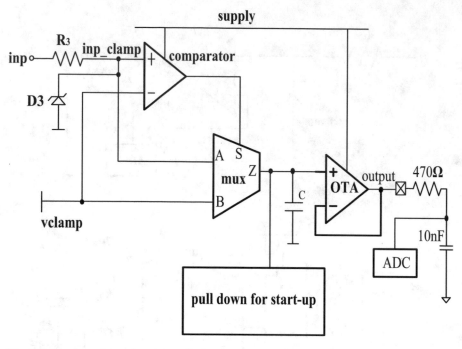

Fig. 6.8 Precise voltage clamping method

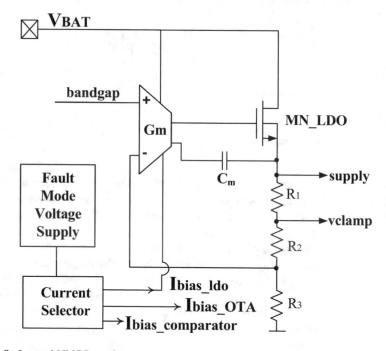

Fig. 6.9 Internal NMOS regulator

Fig. 6.10 Waveforms of
the clamping circuit at
steady state

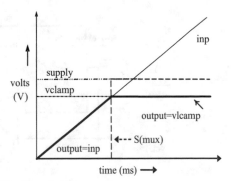

Figure 6.10 shows the waveforms with respect to the steady-state phase of the
OTA. "inp" node refers to the input signal with a huge voltage swing. The swing
will range from 100 mV to 30 V. "Supply" refers to the supply voltage for the
comparator and the OTA. By adjusting the resistor divider, "supply" is set to $\sim$
vclamp + 2 V. R_3 can be adjusted by trim to make vclamp programmable. The
vclamp is chosen to be 5.4 V and is set to 300 mV below the minimum clamping
voltage of the Zener diode D_3. When the comparator output is high, "vclamp" is
mux-ed out to the input of the OTA. When low, the inputs below the "vclamp" level
are mux-ed out to the input of the OTA. The advantage with this approach is that the
noise on V_{BAT} does not propagate to the OTA. The 470 Ω and 10 nF low-pass filter
reduces the noise on the ADC inputs.

Start-Up Clamping Method
During start-up phase, the "V_{BAT}" voltage ramps up from 0 to 40 V. After a few μs,
the bandgap is available ensuring the availability of the "supply" and "vclamp"
levels. The input "inp" to the comparator and the mux is undefined at start-up as the
channels have not been activated by SPI. In this situation the input of the OTA
could rise towards the "supply" level through the C_{gs} of the differential input
transistor pair in the OTA. The "output" node could follow the input of the OTA.
This results in overstressing the ADC during the start-up phase. This can be
overcome with the combination of the transistors M1–M4 and the capacitor C as
shown in Fig. 6.11a.

The transistor M1 is driven by the "power-on reset (POR)" signal. The drain of
M1 is clamped to approximately gate-source voltage (V_{gs}) of M4–V_{th4} where V_{th4} is
the threshold of M4. At start-up the "POR" signal is zero, and so it pulls down the
input of the OTA to zero. The capacitor "C" ensures additional clamping for the fast
transient conditions $\sim$ 10 ns. The voltage behaviours on the various nodes from the
block diagram in Fig. 6.11b.

Simulation results are shown in Fig. 6.12. The "inp" signal is ramped up from
100 mV to 25 V in 10 μs and similarly ramped down from 25 V to 100 mV in 10 μs.
The output signal follows the input signals till 5.4 V, and when the "inp" signal
exceeds 5.4 V, the output is clamped at 5.4 V. The statistical simulation of the
output clamping voltage is shown in Fig. 6.13. The standard deviation "σ" is 2 mV

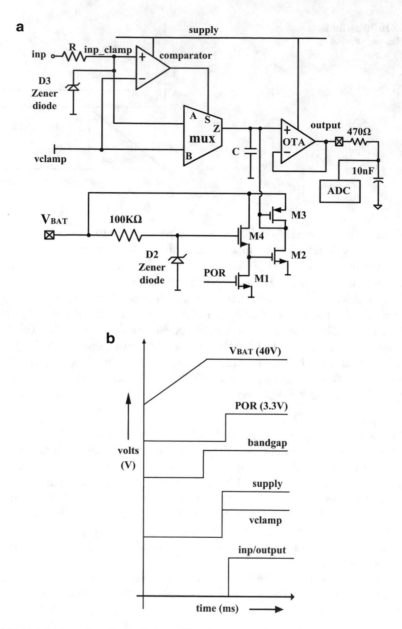

Fig. 6.11 (a) Start-up clamping technique. (b) Voltage on various nodes

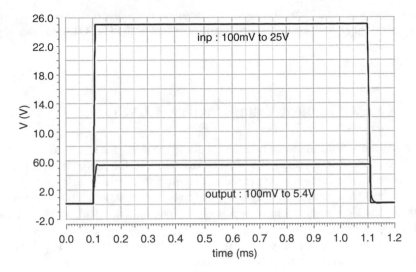

Fig. 6.12 Simulation result showing the input and output signals

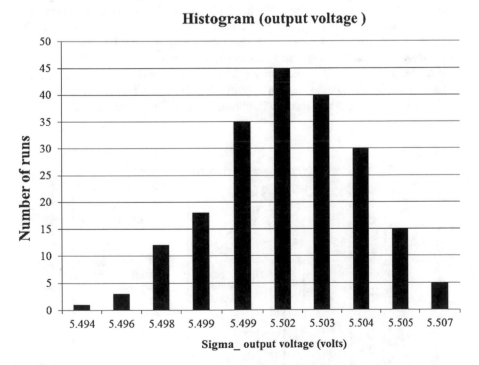

Fig. 6.13 Statistical simulation result of the output clamping level

at a temperature of 175 °C. Higher temperature is the worst-case condition as the leakage is higher, and due to reduced open-loop gain of the HV, OTA and the offset will be higher.

6.3   Current Limited Voltage Source (CLVS)

Current limited voltage source is basically a voltage reference that is critical for the squib resistance measurement (SRM). The resistance of the squib "R" is unknown, and so diagnostic measures are needed to calculate this resistance. This diagnostic circuit needs to be fault tolerant against short to battery or short to ground to prevent high current flow resulting in high power dissipation. The SRM involves a two-point measurement where the current source $I_{src1,2}$ is injected sequentially into the ZMx pin. This is shown in Fig. 6.14. This current flows out of the ZMx pin into the squib resistor and flows back to the Zener diode D_1 and the current sink $I_{snk1,2}$ connected to the Zx pin of the squib driver ASIC. The Zener diode D_1 is needed to set a certain c on Zx for the SRM. $I_{snk1,2}$ is needed to limit the current through D_1 in case of short-to-battery fault on the Zx pin. For instance, I_{src1} flowing through "R" will provide a drop of V_{src1}. Similarly the voltage drop V_{src2} is obtained from I_{src2}.

It is important to have $I_{snk1,2}$ higher than $I_{src1,2}$ so that $I_{src1,2}$ are the dominant currents. If two currents are equal, the regulation loop will not be able to regulate the voltage to the desired target and very likely will settle down to a voltage equal to half of the supply rail Vdd. The voltage drops $V_{src1,2}$ now become the differential input voltage for the instrumentation amplifier [7] OPA with a certain gain "G". The $V_{output1,2}$ of the OPA is measured at the "output" pin. The parameters G, $I_{src1,2}$ are

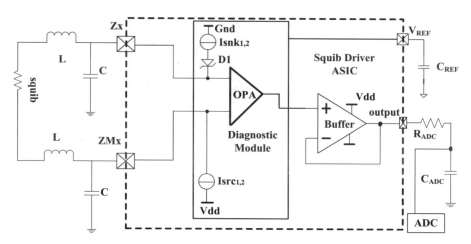

Fig. 6.14 Concept of current limited voltage source (CLVS)

known, and $V_{\text{output1,2}}$ are directly measured. The unknown value of the squib resistance "R" is calculated as per the Eq. (6.1).

Since the range of the squib resistance is between 1 Ω and 8 Ω, the currents $I_{\text{src1,2}}$ are typically in the order of 10–40 mA. Gain of the OPA is also programmable. A low gain value of 10 and a high gain of 20 are typically used in the system. Lower gain is used for high ohmic squibs to prevent railing out of the AMX_OUTPUT. The higher gain is used for low ohmic squibs to ensure a higher differential input voltage.

Known parameters are the gain of the OPA for $I_{\text{src 1,2}}$. We represent this by G. The directly measured parameter is $V_{\text{output1,2}}$. $\Delta V_{\text{src1,2}}$ are calculated from $V_{\text{output1,2}}$ by dividing $V_{\text{output1,2}}$ by the OPA gain G.

$$
R = \left| \frac{\left(\frac{V_{\text{output1}}}{G}\right) - \left(\frac{V_{\text{output2}}}{G}\right)}{I_{\text{src1}} - I_{\text{src2}}} \right| \tag{6.1}
$$

Current Sink and Zener Diode

Since the currents that are used for this measurement is in several tens of mA, on-chip Zener diodes have to be really large to handle these currents so that reliability issues do not occur. This is not area effective, and the on-chip Zener break voltage is not very precise in many process technologies. Hence the current limited voltage source (CLVS) implemented by using a Zener diode will have a wide spread due to the process and temperature variation. This variation is in the order of 1–2 V. To overcome this process, independent voltage reference with current limitation is desired.

This precision can be achieved by regulating the voltage at the output of the HS_FET, i.e. the Zx pin (HS regulation), or at the output or the drain of the LS_FET, i.e. the ZMx pin (LS regulation), by using the voltage regulation loops [7]. The HS regulation-based CLVS architecture is discussed here. Also different current levels for $I_{\text{src1,2}}$ and $I_{\text{snk1,2}}$ have been chosen in order to design a robust regulation loop without any sensitivity to these current levels. Table 6.1 shows the regulation schemes and the magnitudes of the currents that are used here.

Table 6.1 CLVS regulation voltage

Voltage regulation node	Topology name	I_{src1} (mA)	I_{src2} (mA)	I_{snk1} (mA)	I_{snk2} (mA)	Voltage output (volts)
ZMx	High side (HS)	12	36	8	24	8.4 Zx = 8.4 - $(I_{\text{src1,2}}.R)$

6.4   CLVS with High-Side (HS) Regulation

The voltage source with current limitation can be implemented by the conventional NMOS or PMOS LDO-type circuits. Here an LDMOS (NMOS)-type powerFET has been preferred as the PMOS-type powerFET is not efficient in terms of area. The PMOS output stage is an inverting output stage, whereas the NMOS output stage is non-inverting stage. Assuming a two-stage error amplifier that has two inverting stages, the additional inversion with the PMOS-type output can be avoided by using the NMOS output stage in order to simplify the frequency compensation. The load is R-L-C type with the inductance loads at the output ranging from 1 µH to 100 µH, the capacitance from 22 nF to 220 nF and the squib resistance from 1 Ω to 8 Ω. The implemented CLVS with NMOS-type output stage is shown in Fig. 6.15.

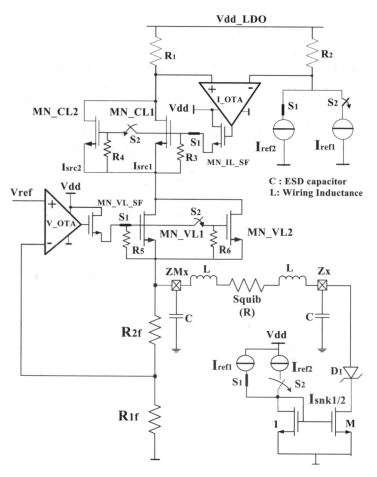

Fig. 6.15 Current limited voltage source with regulated ZMx voltage

The transistors MN_VL1,2 and MN_IL1,2 are the driver FETs for the CLVS output. For analysis let us assume S_1 switch is active and S_2 is inactive. This indicates that the transistor MN_IL1 is connected to I_OTA. Let us analyse the circuit from two load conditions on the ZMx pin. When there is no load on the ZMx pin, the current through the transistor MN_IL1 is lower than I_{ref1}, thereby the output of the amplifier I_OTA swings towards Vdd, thereby fully turning *on* the transistor MN_IL1. This transistor operates in R_{ds_on} mode. The amplifier V_OTA and the transistor MN_VL1 form the traditional NMOS LDO topology, thereby regulating the ZMx pin as shown in the equation below.

As the external load on the ZMx pin would keep increasing, the voltage will get regulated, till the I_{src1} current exceeds I_{LIMIT} value. Further load current increase is not possible, and the amplifier I_OTA will regulate the current to I_{LIMIT}. So I_{src1} equals I_{LIMIT}. MN_VL1 will be operating in R_{ds_on} mode as the amplifier V_OTA will swing towards Vdd. This is the protection against the short-to-ground fault. When I_{snk1} is connected through the squib resistor R and if $I_{src1} > I_{snk1}$, then voltage is regulated to VZMx level, and the I_{src1} current equals the I_{snk1} current by forcing the transistor MN_IL1 into R_{ds_on} mode. If there is a short to the battery, the I_{src1} path is off, and the current is limited to the I_{snk1} value. The same applies when S_2 switch is active. MN_VL2 and MN_CL2 are now activated along with MN_VL1 and MN_CL1.

$$V_{ZMx} = \left(1 + \frac{R_{2f}}{R_{1f}}\right) V_{ref}$$
$$I_{LIMIT} = I_{src1} = \left(\frac{R_2}{R_1}\right) I_{ref1} \qquad (6.2)$$
$$I_{snk1} = M I_{ref1}$$

When S_2 is closed, I_{ref2} is $K I_{ref1}$ where K is 3. This scales I_{LIMIT} (I_{src2}) and I_{snk2} by a factor of k.

$$I_{LIMIT} = I_{src2} = \left(\frac{R_2}{R_1}\right) 3 I_{ref1} \qquad (6.3)$$
$$I_{snk2} = 3M I_{ref1}$$

Voltage Loop Small-Signal Analysis
The small-signal analysis is performed loop by loop. If both V_OTA and I_OTA loops are stable, the complete CLVS circuit is stable. The stability of the amplifier I_OTA is critical only during short-to-ground faults. For SRM topology, the CLVS regulates the voltage, and the amplifier I_OTA is not in the critical path for stability as the amplifier behaves like a comparator. The open-loop analysis is performed on the V_OTA NMOS LDO stage. Let us assume the switch in S_2 is active. Since I_{src2} is lower than I_{snk2}, I_{snk2} is the load current. For voltage mode analysis, the resistance R_1 can be neglected so that the drain of the MN_VLx can be treated as a small-signal ground. Based on this, the small-signal model of the uncompensated V_OTA

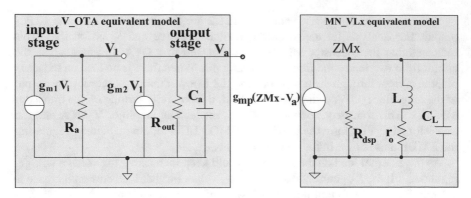

Fig. 6.16 Small-signal model of uncompensated V_OTA with MN_VLx

loop is shown in Fig. 6.16. It is modelled as a two-stage amplifier with only one high impedance node. The output stage is connected to the ZMx node. The transfer function $H(s)$, DC again and the poles can be deduced as.

$$H(s) \sim \frac{g_{m1}R_a g_{m2}R_{out}g_{mp}r_o\left(1+s\dfrac{L}{r_o}\right)}{\left[1+s\left(g_{mp}L+C_aR_{out}\right)+s^2L\left(g_{mp}C_aR_{out}+C_L\right)\right]\left(1+s\dfrac{L}{r_o}\right)}$$

For $L < 100\ \mu H$ and $C_L < 220nF$

$$(6.4)$$

$$g_{mp}L+C_aR_{out} \sim C_aR_{out} \quad \text{and}\ g_{mp}C_aR_{out}+C_L \sim g_{mp}C_aR_{out}$$

$$\Rightarrow H(s) \sim \frac{g_{m1}R_a g_{m2}R_{out}}{\left[1+sC_aR_{out}+s^2\left(\dfrac{C_aC_LR_{out}}{g_{mp}}\right)\right]}$$

$$A_{dc\_2} = g_{m1}R_a g_{m2}R_{out}$$

$$f_{p1\_2} \sim \frac{1}{2\pi C_aR_{out}};$$

$$f_{p2\_2} \sim \frac{g_{mp}}{2\pi C_L} \quad ; \qquad (6.5)$$

$$f_{\tau\_2} = \frac{1}{2\pi}\sqrt{\frac{g_{m1}R_a g_{m2}g_{mp}}{C_L C_a}}$$

For the 24 mA load condition, the poles and zeroes of the uncompensated network are shown below. Assuming the following parameters

$$g_{m1} = 20\mu S, R_a = 20K\Omega, g_{m2} = 30\mu S, R_{out} = 10G\Omega, C_a = 1pF, g_{mp} = 150mS, r_o$$
$$= 350\Omega$$

$$A_{dc\_2} = (20.10^{-6})(20.10^{+3})(30.10^{-6})(10.10^{+9}) = 120000$$

$$\Rightarrow A_{dc\_2} = 20\log_{10}(120000) = 100dB$$

$$f_{p1\_2} \sim \frac{1}{2\pi C_a R_{out}} = \frac{1}{6.282\ (1.10^{-12})(10.10^{+9})} = 15.9\ \text{Hz};$$

$$f_{p2\_2} \sim \frac{g_{mp}}{2\pi C_L} = \frac{150.10^{-3}}{6.282(220.10^{-9})} = 108\ \text{kHz};$$

$$f_{\tau\_2} \sim \frac{g_{m1}R_1g_{m2}g_{mp}}{2\pi C_a C_L} = \frac{(20.10^{-6})(20.10^{+3})(30.10^{-6})(150.10^{-3})}{6.282(1.10^{-12})(220.10^{-9})} = 455\ \text{kHz}$$

The system is unstable due to the presence of the two poles within the UGB. This will result in very low phase margin making the system unstable. The system can be stabilized by moving the first pole to a lower frequency by increasing the value of C_a [8]. Here Miller compensation is chosen in order to stabilize the loop. This is shown in Fig. 6.17. The overall strategy for stability is to move the first pole to a lower frequency and reduce the UGB of the overall loop without altering the second pole and the zero locations. With this approach it is also ensured that for a different I_{snk} load, scaling feature can be applied such that the DC gain, poles, zero and UGB remain unchanged (Fig. 6.17).

Since r_o is 350 Ω, the series impedance of $sL + r_o$ is high when compared to the impedance offered by $1/sC_L$ as the frequency increases. So the impact of the capacitor C_L cannot be ignored. The transfer function $H(s)$ is

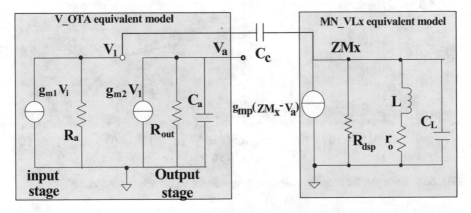

Fig. 6.17 Small -signal model of compensated V_OTA with MN_VLx

$$H(s) \sim \frac{g_{m1}R_a g_{m2}R_{out}}{\left[1 + s g_{m2}R_{out}C_c R_a + s^2\left(\dfrac{C_L C_a R_{out}}{g_{mp}}\right)\right]}$$

$$A_{dc\_2} = g_{m1}R_a g_{m2}R_{out}$$

$$f_{p1\_2} \sim \frac{1}{2\pi g_{m2}R_{out}C_c R_a};$$ (6.6)

$$f_{p2\_2} \sim \frac{g_{mp}g_{m2}R_a C_c}{2\pi C_L C_a}   ;$$

$$f_{\tau\_2} = \frac{g_{m1}}{2\pi C_c}$$

After compensation where the C_c value is increased to 20 pF, the poles and UGB are recalculated Bode plot is shown in Fig. 6.18.

$$g_{m1} = 20\mu S, R_a = 20K\Omega, g_{m2} = 30\mu S, R_{out} = 10G\Omega, C_a = 1pF, C_c = 20pF,\ g_{mp}$$
$$= 150mS,\ r_o = 350\Omega$$

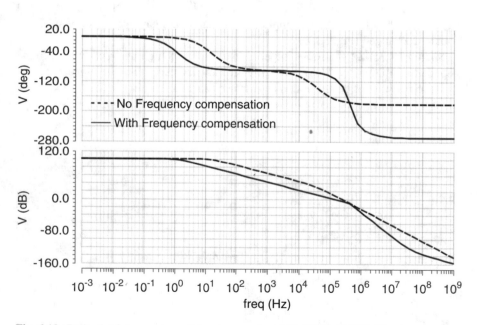

Fig. 6.18 Bode plot for uncompensated and compensated V_OTA with MN_VLx

$$A_{dc\_2} = g_{m1}R_ag_{m2}R_{out} = (20.10^{-6})(20.10^{+3})(30.10^{-6})(10.10^{+9}) = 120000$$
$$\Rightarrow A_{dc\_2} = 20\log_{10}(120000) = 101\ dB$$

$$f_{p1\_2} \sim \frac{1}{2\pi g_{m2}R_{out}C_cR_1} = \frac{1}{6.282(30.10^{-6})(10.10^{+9})(20.10^{-12})(20.10^{+3})} = 1.32\ Hz;$$

$$f_{p2\_2} \sim \frac{g_{mp}g_{m2}R_1C_c}{2\pi C_LC_a} = \frac{(150.10^{-3})(30.10^{-6})(20.10^{+3})(20.10^{-12})}{6.282(220.10^{-9})(1.10^{-12})} = 1.4\ MHz\ ;$$

$$f_{\tau\_2} = \frac{g_{m1}}{2\pi C_c} = \frac{20.10^{-6}}{6.282(1.10^{-12})} = 159\ kHz$$

Phase Margin $\Phi_m = 180^{\circ} - \arctan\left(\dfrac{f_{\tau\_2}}{f_{p1\_2}}\right) - \arctan\left(\dfrac{f_{\tau\_2}}{f_{p2\_2}}\right)$

$$\Rightarrow \Phi_m = 180^{\circ} - \arctan\left(\frac{159K}{1.32}\right) - \arctan\left(\frac{192K}{1.4M}\right) = 90^{\circ}$$

For I_{snk1} setting, S_1 is active and S_2 is *off*. Here the load current is scaled down by a factor of k, where k is 3. The transconductance g_{mp} is scaled by a factor of k since both I_d and W/L are scaled down by k. Since only MN_VL1 is used, C_2 is scaled down by a factor of k. For load current I_{snk2}, let us analyse the transconductance change. The Bode plot for I_{snk1} and I_{snk2} are the same. as shown in Fig. 6.19.

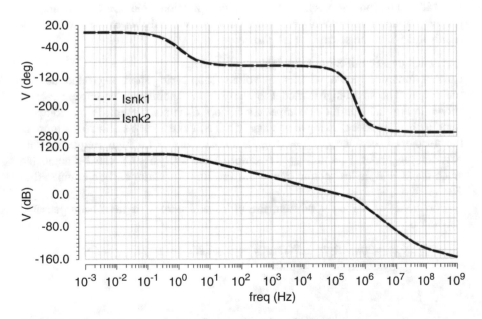

Fig. 6.19 Bode plot for $I_{snk1,2}$ load settings on the voltage loop

For I_{snk2}, transconductance is

$$g_{mp\_snk2} = \sqrt{2I_{d\_snk2}\,\mu C_{ox}\frac{W}{L}}$$

$$g_{mp\_snk1} = \sqrt{2I_{d\_snk1}\,\mu C_{ox}\frac{W}{L}} = \sqrt{2\frac{I_{d\_snk2}}{k}\mu C_{ox}\frac{W}{kL}} = \frac{1}{k}g_{mp\_snk2}$$

$$A_{dc\_1} = g_{m1}R_a g_{m2}R_{out} = g_{m1}R_a g_{m2}R_{out} = A_{dc\_2}$$

$$f_{p1\_1} \sim \frac{1}{2\pi g_{m2}C_c R_{out}R_a} = f_{p1\_2} \;\; ; \;\; f_{p2\_1} \sim \frac{\left(\frac{1}{k}g_{mp\_snk2}\right)C_c g_{m2}R_a}{2\pi C_L \left(\frac{C_a}{k}\right)} = f_{p2\_2} \;\; ;$$

$$f_{\tau\_1} = \frac{g_{m1}}{2\pi C_c} = f_{\tau\_2}$$

$$(6.7)$$

Parameters A_{dc}, f_{p1} and f_{τ} are independent of the load current. The second pole scales accordingly with the load current such that the stability coefficients for I_{snk2} can be made the same as I_{snk1}.

This has several advantages since it reduces the mathematical analysis and SPICE simulations for the different load currents. Advantages are that the mathematical analysis is performed for only one load current. The stability is influenced by L, C, g_m, g_{ds}, temperature, voltage ranges and C_a. For each load, current transconductance (g_{mp}) is different, and the SPICE simulator has to run at least $2^7 \sim 128$ simulation cases which is a huge computation time. On top of this, the process variation and models have to be considered. All this can be confidently eliminated when the scalability is followed.

Current Loop Small-Signal Analysis
Current limitation loop dominates the voltage loop during short-to-ground faults at the output. The small-signal model to analyse the stability of the current loop is shown in Fig. 6.20. In the uncompensated loop, C_c is not present. In the compensated loop, C_c is added taking advantage of the Miller effect to stabilize the loop. In the uncompensated loop, $C_c = 0$ implies that there is no high frequency path between the nodes V_a and V_{out}. To simplify the analysis, C_L can be neglected since the $sL + r_o$ is low ohmic due to current limit conditions when compared to $1/sC_L$.

For the 24 mA load condition, the poles and zeroes of the uncompensated network is

$$g_{m1} = 20\mu S, R_a = 20K\Omega, g_{m2} = 30\mu S, R_{out} = 1G\Omega, C_a = 5pF,$$
$$g_{mp} = 150mS, \; r_o = 350\Omega.$$

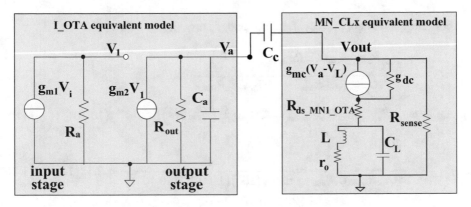

Fig. 6.20 Small-signal model of compensated I_OTA with MN_CLx

For the uncompensated loop, the A_{dc}, f_{p1}, f_{p2} and f_τ are equal to

$$A_{dc\_2} = g_{m1}R_a g_{m2}R_{out}G_{mc}R_{sense}$$

where $G_{mc} = \dfrac{g_{mc}}{1 + g_{mc}R_{ds\_on\_V\_MN1}} \sim \dfrac{1}{R_{ds\_on\_V\_MN1}} = \dfrac{1}{15}$

$$A_{dc\_2} = \left(20.10^{-6}\right)\left(20.10^{+3}\right)\left(30.10^{-6}\right)\left(10.10^{+9}\right)\left(\frac{1}{15}\right)100$$

$$= 800,000 = 20\log_{10}(800000) \sim 118 \text{ dB}$$

$$f_{p1\_2} \sim \frac{1}{2\pi C_a R_{out}} = \frac{1}{6.282\left(1.10^{-12}\right)\left(10.10^{+9}\right)} = 1.59 \text{ Hz};$$

$$f_{p2\_2} \sim \frac{1}{2\pi G_{mc}L} = \frac{1}{6.282\,(15)\left(100.10^{-6}\right)} = 23.8 \text{ kHz};$$

$$f_{\tau\_2} \sim \frac{1}{2\pi}\sqrt{\frac{g_{m1}R_a g_{m2}R_{sense}}{LC_a}} = \frac{1}{2\pi}\sqrt{\frac{\left(20.10^{-6}\right)\left(20.10^{+3}\right)\left(30.10^{-6}\right)\left(10.10^{+9}\right).100}{\left(100.10^{-6}\right)\left(1.10^{-12}\right)}}$$

$$= \frac{3464}{6.282} = 551.4 \text{ kHz}$$

For the stable loop, A_{dc}, f_{p1}, f_{p2} and f_τ are shown below

$$H(s) = \frac{g_{m1}R_a g_{m2}R_{out}G_{mc}R_{sense}\left(1 - \dfrac{sC_c}{G_{mc}}\right)}{\left(1 + sC_c R_{out}G_{mc}R_{sense} + s^2 G_{mc}LC_c R_{out}\right)}$$

$$A_{dc\_2} = g_{m1}R_a g_{m2}R_{out}G_{mc}R_{sense} = \left(20.10^{-6}\right)\left(20.10^{+3}\right)\left(30.10^{-6}\right)\left(10.10^{+9}\right)\left(\frac{1}{15}\right)100$$

$$= 120 \text{ dB}$$

$$f_{p1\_2} \sim \frac{1}{2\pi G_{mc}R_{sense}C_c R_{out}} = \frac{15}{6.282\,(100)\left(1.10^{-12}\right)\left(10.10^{+9}\right)} = 2.38\left(1.10^{-4}\right)\left(1.10^{+3}\right)$$

$$= 238 \text{ mHz};$$

$$f_{p2\_2} \sim \frac{R_{sense}}{2\pi L} = \frac{100}{6.282\left(100.10^{-6}\right)} = 159 \text{ kHz };$$

$$f_{z1\_2} = \frac{G_{mc}}{2\pi C_c} = \frac{1}{6.282\left(15.10^{-12}\right)} = 10.6 \text{ GHz}$$

$$f_{\tau\_2} = \frac{g_{m1}R_a g_{m2}}{2\pi C_c} = \frac{\left(20.10^{-6}\right)\left(20.10^{+3}\right)\left(30.10^{-6}\right)}{6.282\left(15.10^{-12}\right)} = 192 \text{ kHz}$$

Phase Margin Φm is

$$\Phi_m = 180^\circ - arctan\left(\frac{f_{\tau\_2}}{f_{p1\_2}}\right) - arctan\left(\frac{f_{\tau\_2}}{f_{p2\_2}}\right) + arctan\left(\frac{f_{\tau\_2}}{f_{z1\_2}}\right)$$

$$\Rightarrow \Phi_m = 180^\circ - arctan\left(\frac{192K}{0.238}\right) - arctan\left(\frac{192K}{159K}\right) + arctan\left(\frac{192K}{10.6G}\right) = 40^\circ$$

For the low-load conditions, g_{mc} is scaled down by a factor of 3. However, G_{mc} is unaltered since the product $g_{mc}R_{ds\_on\_V\_MN1}$ is higher than 1. For I_{snk1}, G_{mc} is insensitive to load current change. A_{dc1} and A_{dc2} are equal. The poles f_{p1_2}, f_{p2_2}, f_{z1_2} and f_{τ_2} are independent of g_{mc}. Hence the poles, zero and UGB are the same for both I_{snk1} and I_{snk2}. Figure 6.21 shows the Bode plot for the uncompensated and compensated loop.

The Bode plot for the current loop model is shown in Fig. 6.22, and the Bode plot for the implemented circuit is shown in Fig. 6.23. It matches well to the ideal model. However the L-C combination is included in the loads, phase response has a sudden change indicating a complex pole and such a response is typical for a complex R-L-C load. This is also observable in the Bode plot, and design measures should be taken in order to ensure that there is no gain peaking due to the complex pole [8]. Gain peaking close to the UGB will result in degraded phase margin and can create instability in the overall loop. This will reflect as a poor settling response with heavy ringing before settling to its steady-state value.

Additionally the phase margin, over inductance and capacitive load variations are shown in Fig. 6.24. The capacitive load is kept at 22 nF, and the inductance is

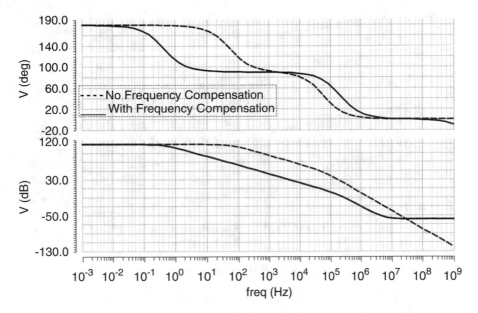

Fig. 6.21 Bode plot for compensated and uncompensated loops – current loop model

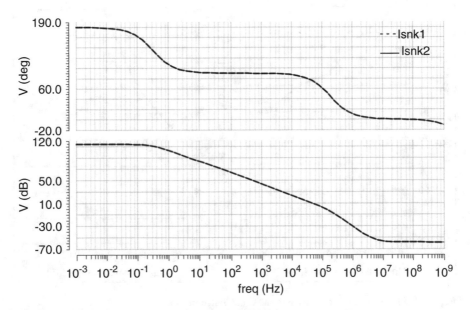

Fig. 6.22 Bode plot for compensated current loop model

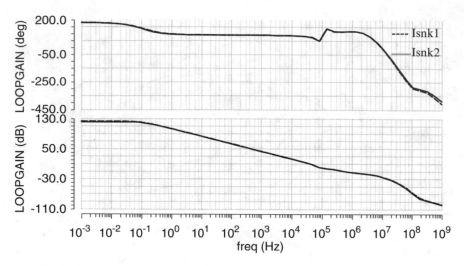

Fig. 6.23 Bode plot for the implemented HS CLVS circuit

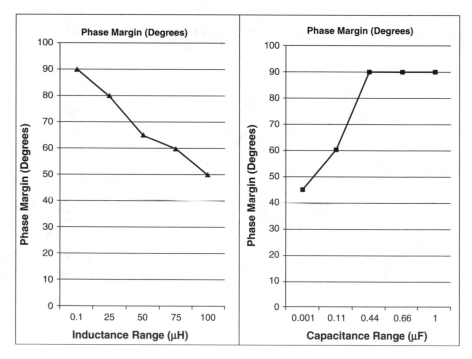

Fig. 6.24 Phase margin for inductance and capacitance ranges

varied from 100 nH to 100 μH range. The phase margin reduces with higher inductances as the 2nd pole f_{p2} moves towards low-frequency region. Similarly inductance is set to 100 μH, and the phase margin is checked for the capacitance

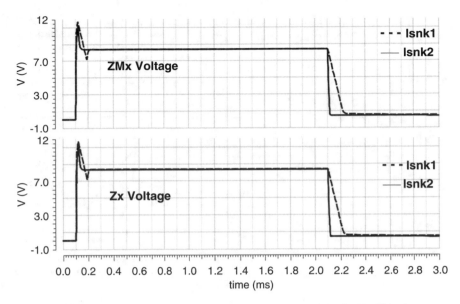

Fig. 6.25 Transient response showing voltage regulation with 22 nF and 100 μH load

range from 100 pF to 1 μF. The phase margin is 90° indicating that capacitance is not influencing the stability of the current loop significantly.

With L being in μH range and capacitance in nF range, the R-L path influences the stability significantly than capacitive path. The transient response for the high-side CLVS is shown in Figs. 6.25 and 6.26 for 22 nF and 220 nF capacitive loads. The voltage overshoot at start-up is due to the in-rush current charging the output capacitor. This voltage peak is limited by the current limitation loop.

6.5   Measurement Results

The quiescent current measurement results during phase I to III are shown in Fig. 6.27. The HV and MV rails were ramped up and ramped down as shown. Based on thermal simulations, the ambient temperature was set to 150 °C so that the on-chip junction temperature is 175 °C. The quiescent current was measured to be 4.1 mA indicating small leakage at high temperature. No high current due to unintended activation or extremely low current due to unintended deactivation was observed. Internal probe pads were accessed at room temperature to measure the voltage and current selector output MAVS and I_{OUT_MCS1}. 15 V was forced on I_{OUT_MCS1} pad to measure the current.

Figure 6.28a demonstrates the different phases of operation, ensuring that the LS driver is not activated inadvertently or deactivated inadvertently. Figure 6.28b is a zoomed version of Fig. 6.28a showing that the 3A current limitation of the LS

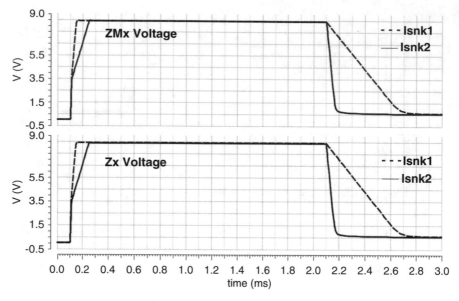

Fig. 6.26 Transient response showing voltage regulation with 220 nF and 100 μH load

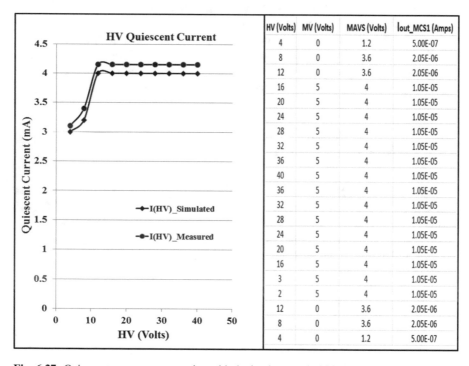

Fig. 6.27 Quiescent current consumption with the implemented MCS-FMVS circuit

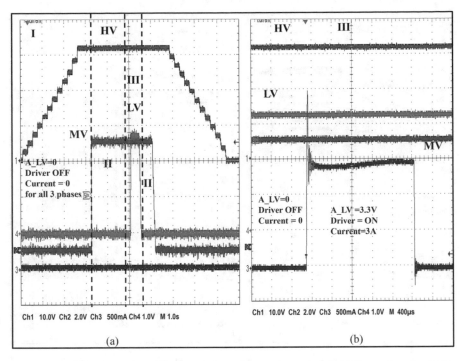

Fig. 6.28 (**a**) Behaviour of the LS driver during phase I to III operation. (**b**) Zoomed version of (**a**) showing LS driver current limitation

driver works only during phase III where all three levels are present and when the signal A_LV is high.

Figure 6.29 shows the output clamping level when the input level changes from 100 mV to 30 V. The result is the same for both the rising and falling edges. Figure 6.30 shows the statistical distribution of the HV OTA. It is within ±100 mV range. Figure 6.31 shows the measured voltage regulation of the CLVS with the HS architecture. Voltage is regulated to 8.4 V. The current loop limits the current and charges the output capacitor, thereby charging the output voltage from 0 V to 8.4 V. The rising edge is symmetrical for both the I_{src1} and I_{src2} settings. Falling edge behaviour is asymmetrical since it is determined by the output capacitor and the load I_{snk1} or I_{snk2}. Since I_{snk2} is 36 mA and is higher than I_{snk1} 4 mA, the falling edge is faster approximately nine times faster for I_{src2} than I_{src1} setting.

6.6   Chip Layout

The layout of the quad channel squib driver is shown in Fig. 6.32. The different modules present in the driver are highlighted. The drivers comprising of the HS_FET and the LS_FET are placed at the corners of the chip to aid easy routing

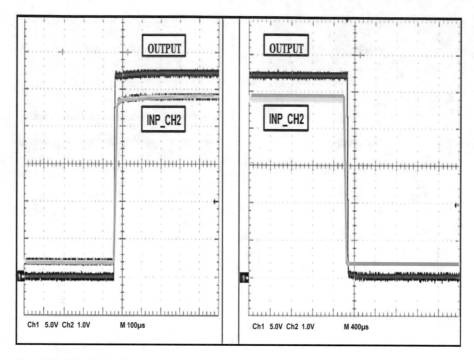

Fig. 6.29 HV OTA with voltage limitation for rising and falling edges of the input

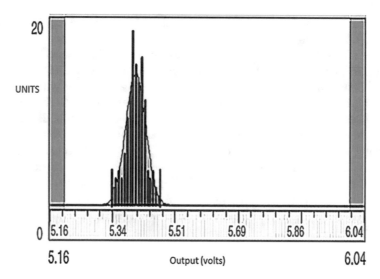

Fig. 6.30 HV OTA voltage limitation precision histogram

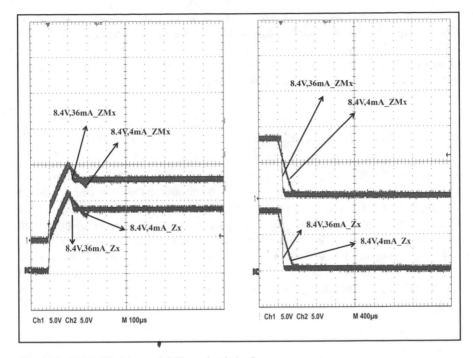

Fig. 6.31 CLVS HS rising and falling edge behaviour

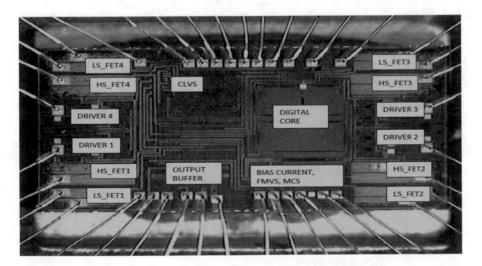

Fig. 6.32 Quad channel squib driver unit chip layout

and minimizing the routing resistance. The diagnostic modules containing the
CLVS output buffer is placed in the middle of the chip. It occupies the entire
mid-area of the chip, mainly due to the need for four switches to activate each
channel independently. The digital core, power supply module with current and
voltage selectors are placed after the diagnostic circuitry.

6.7 Summary and Conclusions

This summarizes the design approach used in ensuring a clean biasing for the quad
channel squib driver ensuring that the tri-stated conditions are avoided. Once the
biasing is well defined, the design techniques ensure that the output voltage of the
buffer is limited to 5.4 V; thereby the low-voltage circuits like the ADCs connected
to the buffer are protected from high voltages. The critical voltage regulation
required during the squib resistance measurement is explained, and the current
limited voltage source required for the diagnostics is presented. The design of
CLVS by using HS-based regulation is presented.

References

1. M.H. Kim, Level shifter having single voltage source, U.S. Patent US7,683,667B2, 23 Mar
   2010
2. C.Y. Chen, System and method for breakdown protection in start-up sequence with multiple
   power domains, U.S. Patent 7391595B2, 24 June 2008
3. B. Sharma et al., Avoiding excessive cross-terminal voltages of low voltage transistors due to
   undesirable supply-sequencing in environments with higher supply voltages, U.S. Patent
   US7,176,749B2
4. Y. Date et al., Current driver, U.S. Patent 7,466,166B2, Dec 2008
5. S.N. Easwaran, I. Hehemann, Bias current generator for multiple supply voltage circuit, U.S.
   Patent US7,888,993B2, 15 Feb 2011
6. S.N. Easwaran, S.V. Kashyap, R. Weigel, Voltage clamping circuit with ±100mV precision in
   high voltage OTA, *Electronic Letters*, 2pp, Online ISSN 1350-911X, published 12 Apr 2016
7. B. Razavi, *Design of Analog CMOS Integrated Circuits* (McGraw-Hill, New York, 2001)
8. S.N Easwaran, *A Low Power NMOS LDO in the Philips C050PMU Process*, Master Thesis,
   University of Twente, The Netherlands, May 2006

Chapter 7
Conclusions and Future Work

In Chap. 1, airbag squib drivers were introduced, and the specifications of an airbag squib driver were discussed. The current sensing technique along with the regulation methods for HS and LS drivers were discussed. In this chapter, the need to limit the current when short to battery fault occurs in the unpowered state of the ASIC is discussed. It is important to ensure that inadvertent deployment during this fault is avoided.

In Chap. 2, the state-of-the-art current sensing circuits were discussed. It compares the pros and cons of the sense resistor-based current sensing versus senseFET-based current sensing along with the regulation loop. Biasing topologies to avoid tri-stated behaviour was discussed by explaining the issue with simple level shifters. The diagnostic section involving current limited voltage source was discussed. This chapter ends by discussing the electrical specifications of the squib driver and explaining the research objective.

In Chap. 3, the fundamental principles of current sensing are presented. The tri-stated nature of circuits during different phases of operation of the ASIC starting with simple cross-coupled LV to MV level shifters was presented. The lack of bias currents due to the cross-coupled level shifter in a V to I converter was presented, and this issue was extended to LV to HV level shifters. The elementary circuits for automotive design like the maximum and minimum voltage and current selector circuits were discussed. The differential voltage and current amplifier topologies were discussed.

In Chap. 4, the HS current regulation scheme was discussed. The implementation with a metal sense resistor and its layout was presented. The self-heating of the metal resistor while handling high currents and its impact to the current regulation was discussed along with the compensation scheme. Small signal analysis was presented. Thermal simulations were discussed in order to understand the high junction temperature rise during deployment so that an optimal floor plan could be achieved. Simulation and measurement results were presented for under- and overdamped loads. The freewheeling path for the high inductive loads and internal

S.N. Easwaran, *Current Sensing Techniques and Biasing Methods for Smart Power Drivers*, https://doi.org/10.1007/978-3-319-71982-5_7

freewheeling concept to avoid any external Schottky diode was presented. Energy limitation during the unpowered state was presented.

In Chap. 5, the LS current limitation scheme was discussed. A senseFET-based current sensing scheme and current limitation to protect the LS_FET during short to battery fault were discussed. Small signal analysis is presented in order to stabilize the loop effectively. Energy limitation during the unpowered state was presented.

In Chap. 6, the biasing scheme involving the maximum current and voltage selector was presented. Accurate voltage clamping technique for the output buffers was discussed. Current limited voltage source involving the HS-type regulation loops was discussed.

In Chap. 7, the summary is presented in a nut shell.

7.1   Future Work

The biggest challenges for automotive ICs start with the thermal challenges and fault conditions. In the metal sense resistor-based current sensing scheme for the HS_FET, the self-heating effect of the metal resistor is compensated by increasing the reference current to achieve the desired target. This approach was needed since the reference resistor is sized such that the self-heating is negligible. Even though this does not substantially increase the power dissipation, compensation through the self-heating of the reference resistor might be an alternate solution. If the reference resistor is sized slightly narrower such that self-heating dominates, its resistance will increase such that the ratio between reference and sense resistor does not reduce significantly when compared to the ratio without self-heating; the targeted deployment current without reference current compensation can be achieved.

Since we discuss about high temperature rise, designers have to intuitively design the current regulation loops by taking into account the thermal behaviour. SPICE simulation models are limited to 200 °C. Today the self-heating of the metal resistor is not modelled. SPICE simulations cannot be performed beyond 200 °C thereby restricting the designers the placement of the thermally sensitive components to achieve an optimized layout. Some development in this area is necessary; else the designers will have to still learn unknowns from the first silicon and can only get the correct result in the second version of the silicon.

For senseFET-based current sensing, with LDMOS transistors, V_{th} mismatch models may not obey Pelgrom's law and hence will need wider trim range of the reference current to achieve the targeted current limitation under any process technology. More work across the semiconductor industry would be needed to mathematically model LDMOS V_{th} mismatch like Pelgrom's law.

The quiescent current requirement for the Airbag SBC if targeted towards 1 mA or lower implementing the level shifters would need bias currents in the order of 100 nA. This would be challenging as these currents are comparable to the leakage currents at high junction temperatures like 175 °C and above. Hence additional circuits might have to be added in order to ensure long-term reliability and robustness. Current limited voltage source for diagnostics is limited to less than 1 μF external capacitive load; else the phase margin degrades. Since the reference needs to be precise and stable, the phase margin improvements up to 10 μF range are recommended. The airbag market being commoditized demands innovation in terms of smart designs like further optimizing the powerFET area and optimal design of on-chip high-voltage capacitive multipliers, etc. to reduce the area of the chip so that the developments can be performed with lower costs.

Appendix A

Small-Signal Analysis: HS Current Regulation

C_1 and C_d can be neglected to simplify the equations. The transfer function is obtained through nodal analysis:

$$@\text{Node } V_1$$

$$g_{m1}V_i + \frac{V_1}{R_a} = 0 \Rightarrow V_1 = -g_{m1}R_a \qquad \text{(A.1)}$$

$$@\text{Node } V_o$$

$$g_{m2}V_1 + \frac{V_o}{R_{out}} + \frac{V_o}{R_z + \frac{1}{sC_c}} = 0 \quad \Rightarrow g_{m2}V_1 + \frac{V_o}{R_{out}} + \frac{sC_cV_o}{sC_cR_z + 1} = 0$$

$$\Rightarrow g_{m2}V_1R_{out}(sC_cR_z + 1) + V_o(sC_cR_z + 1) + sC_cR_{out}V_o = 0$$

$$\Rightarrow sg_{m2}V_1R_{out}C_cR_z + g_{m2}V_1R_{out} + sC_cR_zV_o + V_o + sC_cR_{out}V_o = 0$$

$$\Rightarrow V_1(g_{m2}R_{out} + sg_{m2}R_{out}C_cR_z) + V_o(1 + sC_cR_{out}) = 0 \quad (\because sC_cR_{out} >> sC_cR_z)$$

$$\Rightarrow \frac{V_o}{V_1} = -\frac{g_{m2}R_{out} + sg_{m2}R_{out}C_cR_z}{1 + sC_cR_{out}} = -\frac{g_{m2}R_{out}(1 + sC_cR_z)}{1 + sC_cR_{out}}$$

$$\text{(A.2)}$$

© Springer International Publishing AG 2018
S.N. Easwaran, *Current Sensing Techniques and Biasing Methods
for Smart Power Drivers*, https://doi.org/10.1007/978-3-319-71982-5

@Node V_f

$$g_{mf}(V_o - V_f) - \frac{V_f}{R_{dsf}} = \frac{(V_f - Z_x)}{R_{gs}}$$

$$\frac{[g_{mf}(V_o - V_f)R_{dsf}] - V_f}{R_{dsf}} = \frac{(V_f - Z_x)}{R_{gs}} \tag{A.3}$$

$$\Rightarrow g_{mf}(V_o - V_f)R_{dsf}R_{gs} - V_f R_{gs} = V_f R_{dsf} - Z_x R_{dsf}$$

$$\Rightarrow g_{mf}R_{dsf}R_{gs}V_o = g_{mf}R_{dsf}R_{gs}V_f + R_{gs}V_f + R_{dsf}V_f - Z_x R_{dsf}$$

Now we analyse the source follower and the output stage.

Assuming that the source follower has unity gain to simplify our calculations, we replace V_o with V_f:

$$g_{mf}R_{dsf}R_{gs}V_f = g_{mf}R_{dsf}R_{gs}V_f + R_{gs}V_f + R_{dsf}V_f - Z_x R_{dsf}$$

With $R_{dsf} \gg R_{gs}$ we get $\quad R_{dsf}V_f - Z_x R_{dsf} = 0 \Rightarrow V_f = Z_x$

@Node Z_x, $\quad g_{mp}(V_f - Z_x) + (V_{sense} - Z_x)g_{dp} = \dfrac{Z_x}{R + sL}$ $\qquad$ (A.4)

@Node V_{sense}, $\quad g_{mp}(V_f - Z_x) + (V_{sense} - Z_x)g_{dp} + \dfrac{V_{sense}}{R_{sense}} = 0$

$$\Rightarrow \frac{Z_x}{R + sL} + \frac{V_{sense}}{R_{sense}} = 0 \Rightarrow \frac{V_{sense}}{Z_x} = -\frac{R_{sense}}{R + sL}$$

$$H(s) = \frac{V_{sense}}{V_i} = \left(\frac{V_{sense}}{Z_x}\right)\left(\frac{Z_x}{V_o}\right)\left(\frac{V_o}{V_1}\right)\left(\frac{V_1}{V_i}\right)$$

$$\Rightarrow H(s) = \left(\frac{R_{sense}}{r_o + sL}\right)(1)\left(\frac{g_{m2}R_{out}(1 + sC_cR_{z1})}{(1 + sC_cR_{out})}\right)(g_{m1}R_a)$$

$$\Rightarrow H(s) = \frac{g_{m1}R_a g_{m2}R_{out}R_{sense}(1 + sC_cR_{z1})}{r_o\left(1 + sC_cR_{out} + s^2\dfrac{C_cR_{out}L}{r_o}\right)}$$

$$A_{dc} = \frac{g_{m1}R_a g_{m2}R_{out}R_{sense}}{r_o} \tag{A.5}$$

$$1 + sC_cR_{out} + s^2\frac{C_cR_{out}L}{r_o} = 1 + \frac{s}{p_1} + \frac{s^2}{p_1 p_2}$$

Comparing both we get $p_1 = \dfrac{1}{C_cR_{out}}$ and $p_2 = \dfrac{r_o}{L}$

Equating $1 + sC_cR_{z1} = 0$, zero $Z_1 = \dfrac{1}{C_cR_{z1}}$

$$\omega_\tau = A_{dc}p_1 = \frac{g_{m1}R_a g_{m2}R_{out}R_{sense}}{r_o}\frac{1}{C_cR_{out}} = \frac{g_{m1}R_a g_{m2}R_{sense}}{r_o C_c}$$

Appendix B

Small-Signal Analysis: LS Current Regulation

The V_{ds} equalizer is not analysed along with the main loop in order to simplify the calculations.

Assumption, powerFET $C_{gd} = C_{gs} = C_a$. The transfer function is obtained through nodal analysis.

@Node V_g

$$g_{mi}V_a + \frac{(V_a - V_g)}{R_{di}} = \frac{V_g}{R_a} + sV_gC_a + s(V_g - ZMx)C_a$$

$$\Rightarrow g_{mi}R_{di}R_aV_a + V_aR_a - V_gR_a = V_gR_{di} + s(2V_g\ C_aR_{di}R_a) - sZMxC_aR_{di}R_a \quad (B.1)$$

Since $g_{mi}R_{di}R_a \gg R_a$ and $R_a \gg R_{di}$, Eq. (B.1) can be modified as

$$g_{mi}R_{di}R_aV_a - V_gR_a = s(2V_gC_aR_{di}R_a) - sZMxC_aR_{di}R_a \quad (B.2)$$

© Springer International Publishing AG 2018
S.N. Easwaran, *Current Sensing Techniques and Biasing Methods for Smart Power Drivers*, https://doi.org/10.1007/978-3-319-71982-5

@Node ZMx

$$\left(V_g - ZMx\right)sC_a = g_{mp}V_g + ZMxg_{dp} + \frac{ZMx}{r_o + sL}$$

$$\Rightarrow sV_gC_a - sZMxC_a = \frac{g_{mp}(r_o + sL)V_g + ZMx(r_o + sL)g_{dp} + ZMx}{r_o + sL}$$

$$\Rightarrow sV_gC_a(r_o + sL) - sZMxC_a(r_o + sL) = g_{mp}(r_o + sL)Vg + ZMx(r_o + sL)g_{dp}$$

$$\Rightarrow sVgC_ar_o - sZMxC_ar_o = g_{mp}r_oV_g + sg_{mp}r_oLV_g + ZMx(r_o + sL)g_{dp}$$

$$\Rightarrow ZMx = \frac{\left(sr_oC_a - g_{mp}r_o - sg_{mp}L\right)}{g_{dp}r_o + sg_{dp}L}V_g$$

$$\left(\because C_ar_o \ll g_{dp}L\right) \text{ and } ZMxr_og_{dp} + ZMx \sim ZMxr_og_{dp}$$

$$(B.3)$$

Substitute Eq. (B.3) in Eq. (B.1).

$$g_{mi}R_{di}R_aV_a - V_gR_a = s\left(2V_gC_aR_{di}R_a\right) - s\left(\frac{sr_oC_a - g_{mp}r_o - sg_{mp}L}{g_{dp}r_o + sg_{dp}L}\right)V_gC_aR_{di}R_a$$

$$\Rightarrow g_{mi}R_{di}R_aV_a\left(g_{dp}r_o + sg_{dp}L\right)$$

$$= V_g\left(\begin{array}{c} R_a\left(g_{dp}r_o + sg_{dp}L\right) + 2sC_aR_{di}R_a\left(g_{dp}r_o + sg_{dp}L\right) \\ -s\left(sr_oC_a - g_{mp}r_o - sg_{mp}L\right)C_aR_{di}R_a \end{array}\right)$$

$$\Rightarrow g_{mi}R_{di}R_aV_a\left(g_{dp}r_o + sg_{dp}L\right)$$

$$= V_g\left(\begin{array}{c} g_{dp}r_oR_a + s\left(g_{dp}R_aL + C_aR_{di}R_ag_{mp}r_o + 2C_aR_{di}R_ag_{dp}r_o\right) \\ +s^2\left(2C_aR_{di}R_ag_{dp}L + r_oC_a^2R_{di}R_a + g_{mp}LC_aR_{di}R_a\right) \end{array}\right)$$

For the powerFET g_{mp} and g_{dp} are comparable. Hence $g_{mp} = g_{dp}$.

$$\Rightarrow g_{dp}R_aL + 2C_aR_{di}g_{dp}r_oR_ag_{dp}r_o + g_{mp}C_aR_{di}r_o \sim 3C_aR_{di}R_ag_{mp}r_o;$$

$$\Rightarrow 2g_{dp}LC_aR_{di}R_a + r_oC_a^2R_{di}R_a + g_{mp}LC_aR_{di}R_a \sim 3g_{mp}LC_aR_{di}R_a;$$

Cancelling g_{dp} and R_a, the equation can be simplified to

$$g_{mi}R_{di}\, r_o V_a \left(1 + \frac{sL}{r_o}\right) = V_g \left(g_{dp} r_o + 3 s C_a R_{di} g_{mp} r_o + s^2 \left(3 g_{mp} L C_a R_{di}\right)\right)$$

$$\Rightarrow \frac{V_g}{V_a} = \frac{g_{mi}R_{di} r_o \left(1 + \dfrac{sL}{r_o}\right)}{g_{dp} r_o \left(1 + s\,\dfrac{3 C_a R_{di} g_{mp} r_o}{g_{dp} r_o} + s^2 \dfrac{3 g_{mp} L C_a R_{di}}{g_{dp} r_o}\right)}$$

$$= \frac{\dfrac{g_{mi}R_{di}}{g_{dp}} \left(1 + \dfrac{sL}{r_o}\right)}{\left(1 + s\,\dfrac{3 C_a R_{di} g_{mp}}{g_{dp}} + s^2 \dfrac{3 g_{mp} L C_a R_{di}}{g_{dp} r_o}\right)}$$

Gain from V_0 to V_g is $g_{msn} R_2$

$$(B.4)$$

Now the overall transfer function is derived as

$$H(s) \sim \frac{V_o}{V_a} = \frac{V_o}{V_g}\,\frac{V_g}{V_a}$$

$$\Rightarrow \frac{V_o}{V_a} = \frac{\left[\dfrac{g_{mi}R_{di}}{g_{dp}} \left(1 + \dfrac{sL}{r_o}\right)\right] g_{msn} R_2}{\left(1 + s\dfrac{3 C_a R_{di} g_{mp}}{g_{dp}} + s^2 \dfrac{3 g_{mp} L C_a R_{di}}{g_{dp} r_o}\right)} = \frac{\left[\dfrac{g_{mi}R_{di}}{g_{dp}} \left(1 + \dfrac{sL}{r_o}\right)\right] g_{msn} R_2}{\left(1 + s\dfrac{3 C_a}{g_{di}} + s^2 \dfrac{3 L C_a}{g_{di} r_o}\right)} \qquad (B.5)$$

From Eq. (B.5), the DC gain, poles, zeros, and UGB are deduced.

$$A_{dc} = \frac{g_{mi}R_{di}\, g_{msn} R_2}{g_{dp}}$$

$$\frac{1}{p_1} = \frac{3 C_a R_{di} g_{mp}}{g_{dp}} \Rightarrow p_1 = \frac{g_{dp}}{3 C_a R_{di} g_{mp}} = \frac{g_{di}}{3 C_a}$$

$$\frac{1}{p_1 p_2} = \frac{3 g_{mp} L C_a R_{di}}{g_{dp} r_o} \Rightarrow p_2 = \frac{g_{dp} r_o}{3 g_{mp} L C_a R_{di}\, \dfrac{g_{dp}}{3 C_a R_{di} g_{mp}}} = \frac{r_o}{L}$$

$$z_1 = \frac{r_o}{L}$$

$$\omega_\tau = \frac{g_{mi}R_{di}}{g_{dp}}\, g_{msn} R_2\, \frac{g_{dp}}{3 C_a R_{di} g_{mp}} = \frac{g_{mi} g_{msn} R_2}{3 C_a}$$

Appendix C

Small-Signal Analysis: HS_CLVS Voltage Loop

R_a is a low impedance node and has negligible capacitance.

@Node V_1

$$g_{m1}V_i + \frac{V_1}{R_a} + (V_1 - ZMx)\, sC_c = 0$$

$$\Rightarrow g_{m1}R_aV_i + V_1 + V_1 sR_aC_c - sZMxC_cR_a = 0$$

$$\Rightarrow V_1\,(1 + sR_aC_c) = sZMxC_cR_a - g_{m1}R_aV_i \tag{C.1}$$

$$\Rightarrow V_1 = \frac{sZMxC_cR_a - g_{m1}R_aV_i}{(1 + sR_aC_c)}$$

Substitute Eq. (C.1) in the nodal equation derived at node V_a.

@Node V_a

$$g_{m2}V_1 + \frac{V_a}{R_{out}} + sV_aC_a = 0$$

$$\Rightarrow g_{m2}R_{out}V_1 + V_a + sV_aR_{out}C_a = 0$$

$$\Rightarrow g_{m2}R_{out}V1 = -V_a\,(1 + sR_{out}C_a)$$

$$\Rightarrow V_a = \frac{g_{m2}R_{out}\,(g_{m1}R_aV_i - sZMxC_cR_a)}{(1 + sR_{out}C_a)(1 + sR_aC_c)} \tag{C.2}$$

$$\Rightarrow V_a = \frac{g_{m2}R_{out}(g_{m1}R_aV_i - sZMxC_cR_a)}{(1 + sR_{out}C_a + sR_aC_c + s^2R_aR_{out}C_cC_a)}$$

s^2 terms are neglected and $R_{out}C_a \gg R_aC_c$

$$\Rightarrow V_a = \frac{g_{m2}R_{out}\,(g_{m1}R_aV_i - sZMxC_cR_a)}{(1 + sR_{out}C_a)}$$

© Springer International Publishing AG 2018
S.N. Easwaran, *Current Sensing Techniques and Biasing Methods
for Smart Power Drivers*, https://doi.org/10.1007/978-3-319-71982-5

@Node ZMx

$$(V_1 - \text{ZMx})sC_c = g_{mp}(\text{ZMx} - V_a) + \frac{(\text{ZMx} - V_o)}{sL + r_o} + s\text{ZMx}C_L$$

$$\Rightarrow (V_1 - \text{ZMx})sC_c(sL + r_o) = (\text{ZMx} - V_a)\ g_{mp}(sL + r_o) + (\text{ZMx} - V_a)$$

$$+ s\text{ZMx}C_L(sL + r_o)$$

Since $sL + r_o \sim r_o$,

$$(V_1 - \text{ZMx})sC_c r_o = (\text{ZMx} - V_a)\ g_{mp}r_o + (\text{ZMx} - V_a) + s\text{ZMx}C_L r_o$$

$$\Rightarrow V_1 sC_c r_o - s\text{ZMx}C_c r_o = g_{mp}r_o \text{ZMx} - g_{mp}r_o V_a + \text{ZMx} - V_a + s\text{ZMx}C_L r_o$$

$$\Rightarrow V_1 sC_c r_o + g_{mp}r_o V_a + V_a = \text{ZMx}\left(g_{mp}r_o + 1 + sC_L r_o\right)$$

$$\Rightarrow V_1 sC_c r_o + g_{mp}r_o\left(\frac{g_{m2}R_{out}(g_{m1}R_a V_i - s\text{ZMx}C_c R_a)}{1 + sR_{out}C_a}\right) = \text{ZMx}\left(g_{mp}r_o + sC_L r_o\right)$$

$$\Rightarrow V_1 sC_c r_o(1 + sR_{out}C_a) + g_{mp}r_o g_{m2}R_{out}g_{m1}R_a V_i - s\text{ZMx}C_c R_a g_{mp}r_o g_{m2}R_{out}$$

$$= \text{ZMx}\left(g_{mp}r_o + sC_L r_o\right)(1 + sR_{out}C_a)$$

$$\Rightarrow V_1 sC_c r_o + s^2 C_c r_o R_{out}C_a V_1 + g_{mp}r_o g_{m2}R_{out}g_{m1}R_a V_i$$

$$= \text{ZMx}\left[\left(g_{mp}r_o + sC_L r_o\right)(1 + sR_{out}C_a) + sC_c R_a g_{mp}r_o g_{m2}R_{out}\right]$$

Substitute V_1 from Eq. (C.1).

$$\left(\frac{s\text{ZMx}C_c R_a - g_{m1}R_a V_i}{1 + sR_a C_c}\right)sC_c r_o + s^2 C_c r_o R_{out}\ Ca\left(\frac{s\text{ZMx}C_c R_a - g_{m1}R_a V_i}{1 + sR_a C_c}\right)$$

$$+ g_{mp}r_o g_{m2}R_{out}g_{m1}R_a V_i = \text{ZMx}\left[\left(g_{mp}r_o + sC_L r_o\right)(1 + sR_{out}C_a) + sC_c R_a g_{mp}r_o g_{mr}R_{out}\right]$$

$$\Rightarrow s^2 \text{ZMx}C_c^2 R_a r_o - sg_{m1}R_a V_i C_c r_o + s^3 C_c^2 C_a r_o R_{out}R_a \text{ZMx}$$

$$- s^2 g_{m1}R_a C_c r_o R_{out}C_a V_i + g_{mp}r_o g_{m2}R_{out}g_{m1}R_a V_i + sg_{mp}r_o g_{m2}R_{out}g_{m1}R_a^2 C_c V_i$$

$$= \text{ZMx}\left[\left(g_{mp}r_o + sC_L r_o\right)(1 + sR_{out}C_a) + sC_c R_a g_{mp}r_o g_{mr}R_{out}\right]$$

Neglecting the terms with C_c^2, $C_c C_a$, s^3, the equation is simplified as

$$g_{mp}r_o g_{m2}R_{out}g_{m1}R_a V_i - sg_{m1}R_a V_i C_c r_o$$

$$= \text{ZMx}\left[\left(g_{mp}r_o + sC_L r_o\right)(1 + sR_{out}C_a) + sC_c R_a g_{mp}r_o g_{mr}R_{out}\right] \tag{C.3}$$

The RHS (right-hand side) of equation (C.3) can be simplified as

$$\text{ZMx}\left[\left(g_{mp}r_o + sC_L r_o\right)\left(1 + sR_{out}C_a\right) + sC_c R_a g_{mp} r_o g_{m2} R_{out}\right]$$

$$\sim \text{ZMx}\left[g_{mp}r_o + sC_L r_o + sg_{mp}r_o R_{out}C_a + s^2 C_L r_o R_{out}C_a + sC_c R_a g_{mp} r_o g_{m2} R_{out}\right]$$

Since $C_c R_a g_{mp} r_o g_{m2} R_{out} >> C_L r_o + g_{mp} r_o R_{out} C_a$, equation can be simplifed as

$$\sim \text{ZMx}\left[g_{mp}r_o + sC_c R_a g_{mp} r_o g_{m2} R_{out} + s^2 C_L r_o R_{out} C_a\right]$$

$$\Rightarrow \frac{\text{ZMx}}{V_i} = \frac{g_{mp}r_o g_{m2} R_{out} g_{m1} R_a - sg_{m1} R_a C_c r_o}{g_{mp}r_o \left(1 + sC_c R_a g_{m2} R_{out} + \dfrac{s^2 C_L C_a R_{out}}{g_{mp}}\right)}$$

$$\Rightarrow \frac{\text{ZMx}}{V_i} = \frac{g_{mp}r_o \left(g_{m2} R_{out} g_{m1} R_a - \dfrac{sg_{m1} R_a C_c r_o}{g_{mp}r_o}\right)}{g_{mp}r_o \left(1 + sC_c R_a g_{m2} R_{out} + \dfrac{s^2 R_{out} C_a C_L}{g_{mp}}\right)}$$

since $\dfrac{sg_{m1} R_a C_c}{g_{mp}} \sim 0$,

$$\frac{\text{ZMx}}{V_i} = \frac{\left(g_{m2} R_{out} g_{m1} R_a\right)}{\left(1 + sC_c R_a g_{m2} R_{out} + \dfrac{s^2 R_{out} C_a C_L}{g_{mp}}\right)}$$

DC gain $A_{dc} = g_{m2} R_{out} g_{m1} R_a$

$$\frac{1}{p_1} = C_c R_a g_{m2} R_{out} \Rightarrow p_1 = \frac{1}{C_c R_a g_{m2} R_{out}}$$

$$\frac{1}{p_1 p_2} = \frac{R_{out} C_a C_L}{g_{mp}} \Rightarrow p_2 = \frac{g_{mp}}{R_{out} C_a C_L \cdot \dfrac{1}{C_c R_a g_{m2} R_{out}}} = \frac{g_{mp} R_a g_{m2} C_c}{C_a C_L}$$

$$\omega_\tau = g_{m2} R_{out} g_{m1} R_a \cdot \frac{1}{C_c R_a g_{m2} R_{out}} = \frac{g_{m1}}{C_c}$$

$$\text{(C.4)}$$

Appendix D

Small-Signal Analysis: HS_CLVS Current Loop

R_a is a low impedance node and has negligible capacitance.

$$@\text{Node } V_1$$

$$g_{m1}V_i + \frac{V_1}{R_a} = 0$$

$$\Rightarrow g_{m1}R_aV_i + V_1 = 0 \tag{D.1}$$

$$\Rightarrow V_1 = -g_{m1}R_aV_i$$

Substitute Eq. (D.1) for V_a.

$$@\text{Node } V_a$$

$$g_{m2}V_1 + \frac{V_a}{R_{out}} + sC_aV_a + (V_a - V_{out})sC_c = 0$$

$$\Rightarrow g_{m2}V_1 + \frac{V_a}{R_{out}} + s(C_a + C_c)V_a - sC_cV_{out} = 0$$

Since $C_c \gg C_a$, $C_a + C_c \sim C_c$ $\qquad\qquad$ (D.2)

$$g_{m2}V_1 + \frac{V_a}{R_{out}} + sC_cV_a - sC_cV_{out} = 0$$

$$\Rightarrow g_{m2}R_{out}V_1 + V_a + sC_cR_{out}V_a - sC_cR_{out}V_{out} = 0$$

$$\Rightarrow V_a = \frac{-g_{m2}R_{out}V_1 + sC_cR_{out}V_{out}}{1 + sC_cR_{out}}$$

© Springer International Publishing AG 2018
S.N. Easwaran, *Current Sensing Techniques and Biasing Methods
for Smart Power Drivers*, https://doi.org/10.1007/978-3-319-71982-5

@Node V_{out}

$$G_{mc}(V_a - V_L) + \frac{V_{out} - V_L}{R_{dc}} + \frac{V_{out}}{R_{sense}} = (V_a - V_{out})sC_c$$

$$\Rightarrow G_{mc}R_{dc}R_{sense}V_a - G_{mc}R_{dc}R_{sense}V_L + V_{out}R_{sense} - V_L R_{sense} + V_{out}R_{dc}$$

$$= (V_a - V_{out})sC_c R_{dc}R_{sense}$$

$$\Rightarrow G_{mc}R_{dc}R_{sense}V_a - sC_c R_{dc}R_{sense}V_a - V_L(G_{mc}R_{sense}R_{dc} + R_{sense})$$

$$= -V_{out}(R_{dc} + R_{sense} + sC_c R_{dc}R_{sense}) \tag{D.3}$$

Since $R_{dc} + R_{sense} \sim R_{dc}$ and $G_{mc}R_{sense}R_{dc} + R_{dc} \sim G_{mc}R_{sense}R_{dc}$, equation (D.3) can be simplified as

$$G_{mc}R_{dc}R_{sense}V_a - sC_c R_{dc}R_{sense}V_a - V_L(G_{mc}R_{sense}R_{dc}) =$$

$$- V_{out}(R_{dc} + sC_c R_{dc}R_{sense}) \Rightarrow V_L(G_{mc}R_{sense}R_{dc}) \tag{D.4}$$

$$-G_{mc}R_{dc}R_{sense}V_a + sC_c R_{dc}R_{sense}V_a = V_{out}(R_{dc} + sC_c R_{dc}R_{sense})$$

@Node V_L

$$G_{mc}(V_a - V_L) + \frac{V_{out} - V_L}{R_{out}} = \frac{V_L}{sL + r_o}$$

$$\Rightarrow G_{mc}R_{dc}V_a - G_{mc}R_{dc}V_L + V_{out} - V_L = \frac{V_L R_{dc}}{sL + r_o}$$

$$\Rightarrow sG_{mc}R_{dc}LV_a + G_{mc}R_{dc}r_o V_a - sG_{mc}R_{dc}LV_L - G_{mc}R_{dc}r_o V_L + V_{out}sL$$

$$+ V_{out}r_o - sLV_L - r_o V_L = V_L R_{out}$$

$$\Rightarrow sG_{mc}R_{dc}LV_a + G_{mc}R_{dc}r_o V_a + V_{out}sL + V_{out}r_o = V_L R_{dc} + sG_{mc}R_{dc}LV_L$$

$$+ G_{mc}R_{dc}r_o V_L + sLV_L + r_o V_L$$

Since $R_{dc} + G_{mc}R_{dc}r_0 + r_0 \sim R_{dc}$ and $G_{mc}R_{dc}L + L \sim G_{mc}R_{dc}L$, we get

$$sG_{mc}R_{dc}LV_a + G_{mc}R_{dc}r_o V_a + sV_{out}L + V_{out}r_o = V_L(R_{dc} + sG_{mc}R_{dc}L)$$

$$\Rightarrow V_L = \frac{sG_{mc}R_{dc}LV_a + G_{mc}R_{dc}r_o V_a + sV_{out}L + V_{out}r_o}{(R_{dc} + sG_{mc}R_{dc}L)} \tag{D.5}$$

Substitute (D.5) in (D.4).

$$\left(\frac{sG_{mc}R_{dc}LV_a + G_{mc}R_{dc}r_oV_a + sV_{out}L + V_{out}r_o}{(R_{dc} + sG_{mc}R_{dc}L)}\right)(G_{mc}R_{sense}R_{dc})$$

$$-(G_{mc}R_{sense}R_{dc})V_a + sC_cR_{dc}R_{sense}V_a$$

$$= V_{out}(R_{dc} + sC_cR_{dc}R_{sense})$$

$$\Rightarrow [(sG_{mc}R_{dc}LV_a + G_{mc}R_{dc}r_oV_a + sV_{out}L + V_{out}r_o)(G_{mc}R_{sense}R_{dc})]$$

$$- (G_{mc}R_{sense}R_{dc})(R_{dc} + sG_{mc}R_{dc}L)V_a + sC_cR_{dc}R_{sense}V_a(R_{dc} + sG_{mc}R_{dc}L)$$

$$= V_{out}(R_{dc} + sC_cR_{dc}R_{sense})(R_{dc} + sG_{mc}R_{dc}L)$$

$$\Rightarrow sG_{mc}{}^2R_{dc}{}^2LR_{sense}V_a + G_{mc}{}^2R_{dc}{}^2r_oR_{sense}V_a$$

$$+ sG_{mc}R_{sense}R_{dc}LV_{out} + G_{mc}R_{sense}r_oR_{dc}V_{out} - G_{mc}R_{sense}R_{dc}{}^2V_a$$

$$- sG_{mc}{}^2R_{sense}R_{dc}{}^2LV_a + sC_cR_{dc}{}^2R_{sense}V_a + s^2C_cLG_{mc}R_{dc}{}^2R_{sense}V_a$$

$$= V_{out}(R_{dc}{}^2 + sR_{dc}{}^2G_{mc}L + sC_cR_{dc}{}^2R_{sense} + s^2G_{mc}C_cLR_{dc}{}^2)$$

Simplifying terms

$$@V_a G_{mc}{}^2R_{dc}{}^2r_oR_{sense}V_a - G_{mc}R_{sense}R_{dc}{}^2V_a \sim -G_{mc}R_{sense}R_{dc}{}^2V_a$$

$$sG_{mc}{}^2R_{dc}{}^2LR_{sense}V_a - sG_{mc}{}^2R_{sense}R_{dc}{}^2LV_a$$

$$+sC_cR_{dc}{}^2R_{sense}V_a \sim sC_cR_{out}{}^2R_{sense}V_a$$

Simplifying terms $@V_{out}$

$$R_{dc}{}^2 - G_{mc}R_{sense}r_oR_{dc} \sim R_{dc}{}^2sR_{dc}{}^2G_{mc}L + sC_cR_{dc}{}^2R_{sense}$$

$$+sC_cLG_{mc}R_{dc}{}^2R_{sense} - sG_{mc}R_{sense}R_{dc}L \sim sC_cLG_{mc}R_{dc}{}^2R_{sense}$$

Neglecting s^2 terms

$$\Rightarrow V_a \left[-G_{mc}R_{sense}R_{dc}{}^2 + sC_cR_{dc}{}^2R_{sense}\right] = V_{out}\left[R_{dc}{}^2 + sC_cLG_{mc}R_{dc}{}^2R_{sense}\right]$$

$$\Rightarrow V_a = \frac{V_{out}\left[R_{dc}{}^2 + sR_{dc}{}^2G_{mc}L\right]}{\left[-G_{mc}R_{sense}R_{dc}{}^2 + sC_cR_{dc}{}^2R_{sense}\right]} = \frac{V_{out}\left[1 + sG_{mc}L\right]}{\left[-G_{mc}R_{sense} + sC_cR_{sense}\right]}$$

$$(D.6)$$

Substitute V_a in Eq. (D.2).

$$\frac{V_{out}(1+sG_{mc}L)}{(-G_{mc}R_{sense}+sC_cR_{sense})}=\frac{-g_{m2}R_{out}V_1+sC_cR_{out}V_{out}}{1+sC_cR_{out}}$$

$$\Rightarrow V_{out}(1+sG_{mc}L)(1+sC_cR_{out})=-g_{m2}R_{out}V_1(-G_{mc}R_{sense}+sC_cR_{sense})$$
$$+sC_cR_{out}V_{out}(-G_{mc}R_{sense}+sC_cR_{sense})$$

$$\Rightarrow V_{out}\left[1+sG_{mc}L+sC_cR_{out}+s^2G_{mc}LC_cR_{out}+sG_{mc}R_{sense}C_cR_{out}-s^2C_c^2R_{sense}R_{out}\right]$$
$$=V_1\left[g_{m2}R_{out}G_{mc}R_{sense}-sC_cR_{sense}g_{m2}R_{out}\right]$$

Simplifying the terms on V_{out} side we get

$$G_{mc}LC_cR_{out}-C_c^2R_{sense}R_{out}\ \sim\ G_{mc}LC_cR_{out}$$

$$G_{mc}L+C_cR_{out}+G_{mc}R_{sense}C_cR_{out}\ \sim\ G_{mc}R_{sense}C_cR_{out}$$

Equation is simplified as

$$V_{out}\left[1+sG_{mc}R_{sense}C_cR_{out}+s^2G_{mc}LC_cR_{out}\right]=\left[g_{m2}R_{out}G_{mc}R_{sense}-sC_cR_{sense}g_{m2}R_{out}\right]$$

$$\Rightarrow \frac{V_{out}}{V_1}=\frac{g_{m2}R_{out}G_{mc}R_{sense}-sC_cR_{sense}g_{m2}R_{out}}{\left[1+sG_{mc}R_{sense}C_cR_{out}+s^2G_{mc}LC_cR_{out}\right]}$$

$$=\frac{g_{m2}R_{out}G_{mc}R_{sense}\left(1-\dfrac{sC_c}{G_{mc}}\right)}{\left[1+sG_{mc}R_{sense}C_cR_{out}+s^2G_{mc}LC_cR_{out}\right]}$$

$$(D.7)$$

Hence the overall transfer function V_{out}/V_i is

$$H(s)\sim\frac{V_{out}}{V_i}=-\frac{g_{m1}R_ag_{m2}R_{out}G_{mc}R_{sense}\left(1-\dfrac{sC_c}{G_{mc}}\right)}{\left[1+sG_{mc}R_{sense}C_cR_{out}+s^2G_{mc}LC_cR_{out}\right]}$$

DC Gain $A_{dc}=g_{m1}R_ag_{m2}R_{out}G_{mc}R_{sense}$

$$\frac{1}{p_1}=G_{mc}R_{sense}C_cR_{out}\Rightarrow p_1=\frac{1}{G_{mc}R_{sense}C_cR_{out}}$$

$$\frac{1}{p_1p_2}=G_{mc}LC_cR_{out}\Rightarrow p_2=\frac{G_{mc}R_{sense}C_cR_{out}}{G_{mc}LC_cR_{out}}=\frac{R_{sense}}{L}$$

$$z_1=\frac{G_{mc}}{C_c}$$

$$\omega_\tau=g_{m1}R_ag_{m2}R_{out}G_{mc}R_{sense}\cdot\frac{1}{G_{mc}R_{sense}C_cR_{out}}=\frac{g_{m1}R_ag_{m2}}{C_c}$$

List of Publications

Granted Patent Publications

1. S.N. Easwaran, I. Hehemann, Bias current generator for multiple supply voltage circuit, U.S. Patent US 7,888,993B2, 15 Feb 2011
2. S.N. Easwaran, I. Hehemann, Current limited voltage source with wide input current range, U.S. Patent US 8,045,317B2, 25 Oct 2011
3. S.N. Easwaran, M. Wendt, Current driver circuit, U.S. Patent US 8,093,925B2, 10 Jan 2012
4. S.N. Easwaran, Electronic device for controlling a current, U.S. Patent US 8,553,388B2, 8 Oct 2013
5. S.N. Easwaran, Driver control apparatus and methods, U.S. Patent US 9,343,898B2, 17 May 2016

Journal Publications

1. S.N. Easwaran, S.V. Kashyap, R. Weigel, Voltage clamping circuit with ±100mV precision in high voltage OTA, *Electronic Letters*, 2pp, Online ISSN 1350-911X, published 12 Apr 2016
2. S.N. Easwaran, S.V. Kashyap, R. Weigel, Current regulator with energy limitation in the unpowered state featuring bipolar discharge path. *BCTM Conference and Proceedings Publication, IEEE*, pp. 82–.85, ISSN 2378-590x, 2016
3. S. Camdzic, P. Wang, S.N. Easwaran, *Automotive IQPC Conference (System Safety) – Empowering Innovation and Trust in Self-Driving and Highly Automated Applications*. 19–21 Sep 2016, Las Vegas

Accepted Publications

1. S.N. Easwaran, S. Chellamuthu, S.V. Kashyap, R. Weigel, Voltage and current selector based biasing topology for multiple supply voltage circuits. IEEE Trans. Circuits Syst II Exp. Briefs, doi: https://doi.org/10.1109/TCSII.2017.2660525

© Springer International Publishing AG 2018

S.N. Easwaran, *Current Sensing Techniques and Biasing Methods for Smart Power Drivers*, https://doi.org/10.1007/978-3-319-71982-5

2. S.N. Easwaran, R. Weigel, 100mV precision 40V tolerant scalable cap free current limited voltage source for wide input current range. *60th IEEE International Midwest Symposium on Circuits and Systems II*. 6–9 Aug 2017, Boston
3. S.N. Easwaran, S. Camdzic, R. Weigel, Thermal simulation aided 98mJ integrated high side and low side drivers design for safety SOCs. *30th IEEE International System-On-Chip Conference*. 5–8 Sep 2017, Munich

Index

© Springer International Publishing AG 2018
S.N. Easwaran, *Current Sensing Techniques and Biasing Methods
for Smart Power Drivers*, https://doi.org/10.1007/978-3-319-71982-5

Printed in the United States
By Bookmasters